The Economics Salmon Aquaculture

G000090762

Trond Bjørndal

Blackwell Scientific Publications
OXFORD LONDON EDINBURGH
BOSTON MELBOURNE

First published 1990

Set by DP Photosetting, Aylesbury, Bucks
Printed in Great Britain by
St Edmundsbury Press Ltd
Bury St Edmunds, Suffolk.

DISTRIBUTORS

Marston Book Services Ltd
PO Box 87
Oxford OX2 0DT
(*Orders*: Tel: 0865 791155
     Fax: 0865 791927
     Telex: 837515)

USA
  Publishers' Business Services
  PO Box 447
  Brookline Village
  Massachusetts 02147
  (*Orders*: Tel: (617) 524-7678)

Canada
  Oxford University Press
  70 Wynford Drive
  Don Mills
  Ontario M3C 1J9
  (*Orders*: Tel: (416) 441-2941)

Australia
  Blackwell Scientific Publications
  (Australia) Pty Ltd
  107 Barry Street
  Carlton, Victoria 3053
  (*Orders*: Tel: (03) 347-0300)

British Library
Cataloguing in Publication Data
Bjørndal, Trond
  The economics of salmon aquaculture.
  1. Aquaculture industries
  I. Title
  338.3'71

  ISBN 0–632–02704–5

Library of Congress
Cataloging in Publication Data
Bjørndal, Trond.
  The economics of salmon aquaculture/
  Trond Bjørndal.
    p. cm.
  Includes bibliographical references.
  ISBN 0–632–02704–5
  1. Fish-culture—Economic aspects.
  2. Salmon—Economic aspects.
  I. Title.
  SH167.S17B57 1990
  338.3'72755—dc20

Contents

Preface   ix

1   Salmon Production   1
   1.1   The production process in aquaculture   1
   1.2   The supply of salmon   5
   1.3   The emergence of salmon aquaculture   9
   Notes   26

2   Markets for salmon   28
   2.1   Salmon consumption   28
   2.2   Future competition   32
   Notes   37

3   The theory of optimal harvesting of farmed fish – an introduction   38
   3.1   A biological model   38
   3.2   Bioeconomic analysis   42
   3.3   The rotation problem   55
   3.4   Optimal harvesting: examples   57
   3.5   Postscript   62
   Notes   62

4   A harvesting model for a fish farm   64
   4.1   Cash flow analysis   64
   4.2   Optimal harvesting   67
   4.3   Production planning   77
   Notes   78

5   Cost of production   80
   5.1   Smolt production in Norway   80
   5.2   Salmon farming in Norway   87
   5.3   Salmon farming in British Columbia   94
   Notes   99

6   Start-up of a fish farm   100

Bibliography   106

Appendix 1: Optimal harvesting – further analysis   109

Appendix 2: Statistical tables   116

Index   119

To my parents,
Judith and Thorvald Bjørndal

Preface

The salmon aquaculture industry has expanded rapidly in the 1980s. The industry originated in Norway and, as a consequence of its successful development, later spread to a number of countries in Europe, the Americas and Australia. Markets for salmon appear to have expanded in proportion with increases in output. Hitherto, the industry has been production oriented. Recently, however, market conditions appear to be changing. In circumstances like these, more attention must be paid to the markets for the product and the economics and management of salmon aquaculture.

This book contains economic analyses of salmon aquaculture. The purpose of the book is that it can be used as a textbook for senior undergraduate courses, but it should also be of use to those who work in the industry and others who are interested in salmon aquaculture. Certain parts of the book require some economic and mathematical background. However, most parts of the book are accessible also to those who lack such a background.

The book is based on lectures on the economics of salmon aquaculture at the Norwegian School of Economics and Business Administration, Simon Fraser University and Oregon State University and my research in this field.

In Chapter 1, salmon production is analysed. The production process in aquaculture is described, followed by an overview of the supply of salmon from both capture fisheries and aquaculture. Special emphasis is put on the emergence of salmon aquaculture. The demand side is brought into the picture in Chapter 2, where markets for salmon are analysed.

The theory of optimal harvesting of farmed fish is analysed in Chapter 3. This chapter requires some knowledge in calculus. However, many readers will be able to grasp the intuition of the analysis without any formal mathematical background. Based on this theoretical analysis, a practical harvesting model for fish farms is developed in Chapter 4. As an example, the optimal harvesting of chinook salmon is considered. Although the example presented is for salmon aquaculture, the model is applicable to any culture.

Cost of production is analysed in Chapter 5. Two cases studies are from Norway (smolt production and salmon farming) and one from British Columbia (salmon farming). All these cases are based on firms that are fully developed and have reached 'normal' production. The start-up of a fish farm is analysed in Chapter 6 with an example for Norway. In this case study, cash flow budgeting is of primary interest.

To a large extent, the economic analyses are based on Norwegian data. There are two reasons for this. First, as Norway is the leading salmon producer in the world, analyses of economic conditions in this country are of interest to salmon producers worldwide. Second, as Norway also is the pioneer in this field, more data are available than for any other producing country.

A number of persons have in various ways assisted me in the writing of this book. The invisible presence of my editor, Mr Richard Yates, is evident throughout the book. The advice of Mr Yates on both language and substance has greatly enhanced the quality of the book.

I thank my colleagues Ragnar Arnason, Rögnvaldur Hannesson, Terry Heaps, Kjell Salvanes, Susan Shaw and Richard Schwindt for comments and suggestions on various parts of the book. Excellent research assistance has been provided by Mabel Tang, Jing Lee, Jorun Andreassen, Liv Holmefjord and Janne R. Stene. The manuscript has through several iterations been typed by Bente Gunnarsen, Kellie Heath and Anne Kristin Wilhelmsen.

Trond Bjørndal

1 Salmon production

There are six commercially important salmon species. One (*Salmo salar*) is native to the Atlantic Ocean, the other five (all genus *Oncorhynchus*) to the Pacific. Raw salmon is supplied by the capture fisheries that harvest wild stocks and fish farms which culture Atlantic salmon and the more valuable of the Pacific species.

In this chapter, salmon production is analysed. The worldwide supply of salmon from both the capture fishery and aquaculture will be considered, as the products from these two sources compete in the market. The emergence of salmon aquaculture is described in some detail, and the production process in aquaculture is analysed and briefly compared to that of capture fisheries.

1.1 THE PRODUCTION PROCESS IN AQUACULTURE

Aquaculture (fish farming, fish culture, mariculture, sea ranching) can be defined as the human cultivation of organisms in water (fresh, brackish or marine). Aquaculture is distinguished from other aquatic production by the degree of human *intervention* and *control* that is possible. It is in principle more similar to forestry and animal husbandry than to traditional capture fisheries. In other words, aquaculture is stock raising rather than hunting.

The production process in aquaculture is determined by biological, technological, economic and environmental factors. Many aspects of the production process can be brought under human control. Environmental conditions can be controlled, genetics manipulated to improve yields and harvesting can be timed to ensure continuous supplies of fresh product. This is in sharp contrast to the capture fisheries which are only controlled through harvesting regulations, if at all. And while search for the resource is a very important part of the production process in capture fisheries, no such effort is required in aquaculture.

As outlined by Shang (1981), a number of criteria can be used to classify an aquaculture system. From an economic point of view the most significant criterion is intensity, i.e. the division into intensive, semi-intensive or extensive forms of culture. Measures of intensity include stocking density, production by area, feeding regimes and input costs. In *intensive* salmon farming, fish are reared

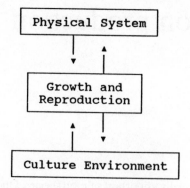

Fig. 1.1 The Production System in Salmon Aquaculture

in pens and the farmer controls factors of production such as farm size, stocking and feeding of fish. Ocean ranching is an *extensive* form of aquaculture in which a body of sea is stocked with fish that feed on natural food. (A number of countries have experimented with ocean ranching of salmon with mixed results.)

The production process in aquaculture can be looked at in terms of interactions between technological and biological factors and the culture environment. This is illustrated in Figure 1.1 (from Allen *et al.*, 1984).

Physical system

Salmon are usually raised in sea pens although there are examples of salmon being raised in closed-off channels. There are two main pen systems in use:

(1) Single pens attached to floaters that are anchored individually. The pens may be of different sizes and configurations.
(2) Platform installations. In this system pens are attached to a float (which may or may not be attached to the shore by a walkway). It is possible to move the whole installation from one location to another. Because of the platform, labour-saving equipment for feeding, harvesting and net cleaning can be used.

Originally, single pens were used for raising salmon. Platform installations were developed in the 1980s and have now become the predominant technology. Commonly, there are land-based facilities for feed production and storage and workers' quarters. However, self-contained platform units with housing and storage facilities placed on a float are becoming more common. Another technology is represented by land-based farms; production is undertaken in tanks or raceways on land with water pumped from the ocean. This technology gives a high degree of control over factors affecting production but due to high investment costs it has not yet become common for grow-out operations.

Feeding methods range from handfeeding to data-controlled automatic feeders. Most farms employ a combination of hand and automatic feeding.

Biological system

Salmon are anadromous fish. Eggs are spawned and hatched in fresh water and the fry remain in fresh water for varying periods. Eventually fry go through a complex physical change known as the smoltification process. During this process they adapt to saltwater life. When it is complete, smolts migrate to sea. After spending one to four years at sea (depending on species), the wild fish return to the river where they were born to spawn. After spawning Pacific salmon always die, Atlantic salmon do not.

Based on the lifecycle of wild salmon, the biological process in salmon aquaculture consists of the following steps:

(1) Production of broodstock and roe.
(2) Production of fry (hatcheries).
(3) Production of smolts.
(4) Production of farmed fish.

Although there are differences between Atlantic and Pacific salmon (discussed below), the production cycle is essentially the same for all species.

The biological process starts with a broodstock. (Originally from wild fish, broodstocks are being domesticated over time.) Eggs are stripped from the female, fertilized and transported to a hatchery. After an incubation period of about two months, yolk-sack larvae are hatched. Feeding starts about one month later. This is a delicate stage of the biological process and there is often a high mortality rate at this time, particularly for Atlantic salmon.

The smoltification process represents an important difference between Pacific and Atlantic salmon. Pacific salmon generally smoltify four to six months after being hatched and weigh 6–8 g; most Atlantic salmon fry smoltify 16 months (called one-year olds) after being hatched, the remainder after another year (two-year olds). One-year old Atlantic salmon smolts weigh around 40 g. The smoltification process is very sensitive to slight environmental changes and high mortality will result if temperature and water salinity are not accurately controlled. Domestic smolts are placed in sea pens after smoltification.

Smolts are sold to specialized grow-out farms where they are raised to marketable size. There are substantial variations in the grow-out periods for the various species. Atlantic salmon may be raised for two years to reach weights of 3–6 kg. Chinook will typically be raised for up to two years and to weights similar to Atlantic salmon. Coho, on the other hand, are raised for only 12 to 16 months to a weight of 2 kg.

Regardless of species, the fish must be harvested before spawning. For Atlantic and chinook salmon this occurs about 28 months after smoltification, while coho mature after only 16 months. However, it should be noted that the time of spawning can vary greatly even for fish of the same yearclass; e.g. a fairly high proportion of chinook mature after one year. (This problem may be reduced over time with further domestication of the species.) There may also be a difference between stocks. Scottish and Irish salmon stocks tend to reach maturity as one-year olds, while this is not common for Norwegian stocks. Although Atlantic salmon do not die after spawning, the quality degradation due to spawning would mean waiting for up to another year before harvesting. This is not practical.

On salmon farms the continual monitoring of fish throughout their life cycle as a means of disease control is crucial. There are two particularly critical stages. The first is the period after new smolts are placed in salt water. The second is prior to sexual maturation. When the fish mature they either die (Pacific salmon) or their quality (Atlantic salmon) is reduced so sharply that they are unfit for consumption. Hence, the fish must be slaughtered prior to maturity.

The production of fish for market can be analysed with fry or smolts as an *input* in the production process. This is illustrated in Figure 1.2 (adapted from Ricker, 1975).

It is important to note that with aquaculture, fish reproduction is a separate activity. In the capture fisheries, quantity harvested will determine the size of the spawning stock. This in turn will influence (future) recruitment to the stock. In salmon aquaculture the link between stock and recruitment is broken.

As indicated in Figure 1.2, growth and natural mortality influence stock development. Both these variables are functions of time. The production of farmed fish takes a long time and throughout this period capital is invested in the stock of fish. Therefore the time of harvesting is important. (This will be analysed in detail in Chapter 3.)

In aquaculture one can control the quality and quantity of smolts and several of the factors that influence growth, e.g. feeding and fertilization. The degree of control is partly determined by the technological system. Accordingly, the farmer can have a significant influence over the biological system.

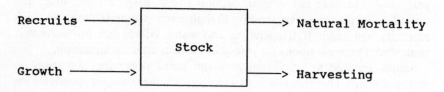

Fig. 1.2 Stock Development

Culture environment

The following culture environment factors influence production:

- flushing action
- temperature
- oxygen
- organic materials
- underwater topography
- protection from winds and waves.

However, after a location is chosen for a sea-pen system these factors can be controlled only to a limited extent. Should the environmental conditions of a site prove detrimental to production, the only solution may be to move the farm to another location. There are several examples of this in Norway. (More of these factors can be brought under human control in land-based farms which are not considered in this book.)

Having considered the production process in aquaculture in conceptual terms, the supply (production) of salmon will now be analysed.

1.2 THE SUPPLY OF SALMON

Raw salmon is supplied by the capture fisheries and fish farms. The relative importance of these two sources is set out in Table 1.1 which shows worldwide landings of the capture fishery and fish farm production for the years 1980–1987.

The total supply of salmon increased markedly through the mid 1980s both as

Table 1.1 Salmon Production and Landings 1980–1987

	Landings of wild salmon ('000 tonnes)	Production of farmed salmon ('000 tonnes)	Total production ('000 tonnes)	Farmed as a per cent of total
1980	573.4	4.8	578.2	1%
1981	649.0	11.6	660.6	2%
1982	557.2	16.5	573.7	3%
1983	678.7	24.6	703.3	3%
1984	624.1	32.6	656.7	5%
1985	793.5	47.1	839.6	6%
1986	675.0	70.0	745.0	9%
1987	650.2	87.5	737.6	12%

Sources: For wild salmon, the Food and Agricultural Organization of the United Nations (FAO). Sources for farmed salmon are listed in Section 1.3

Table 1.2 Average Annual Landings of Salmon for 1980–85 ('000 tonnes)[a]

Atlantic (*Salmo salar*)	11.2
Chinook (*O. tshawytscha*)	21.2
Coho (*O. kisutch*)	35.3
Sockeye (*O. nerka*)	135.7
Pink (*O. gorbuscha*)	238.0
Chum (*O. keta*)	200.9
Cherry (*O. masuo*)	3.6
Total	645.9

[a] It was noted above that there are six commercially important species. A seventh species, cherry, is included here. Annual landings by species are given in Appendix 2, Table A1.

a result of increased wild stock landings (allowances must be made for the cyclical nature of spawning runs; the runs of some species vary greatly from year to year) and a dramatic surge in the salmon aquaculture industry. In 1980 farmed salmon accounted for less than one per cent of all production; by 1987 the figure had risen to 12 per cent. This is a small portion of total production and to fully appreciate the significance of salmon aquaculture in today's production scheme, salmon must be considered on a species by species basis. Table 1.2 gives average annual landings of salmon by species for the period 1980–1985.

Table 1.2 clearly shows that the capture fisheries are dominated by three species, pink, chum and sockeye. Combined they account for 90 per cent of total landings. Catches of coho and chinook are much smaller, while landings of Atlantic and cherry salmon are smaller yet.

Fish farmers specialize in producing three species, Atlantic, coho and chinook. Their output directly competes with the same wild caught species in specific market segments (markets will be discussed in Chapter 2). The importance of salmon aquaculture production of these 'desirable' species is illustrated in Table 1.3. Two alternatives are considered. The first includes

Table 1.3 Production Shares of Farmed Salmon 1980–1985[a]

	Farmed as a per cent of total production of Atlantic, chinook and coho	Farmed as a per cent of total production of Atlantic, chinook, coho and sockeye
1980	7	3
1985	40	18

[a] The production shares are derived from Tables 1.1 and A1 (Appendix 2).

Atlantic, chinook and coho. The second adds sockeye to the list of desirable wild species. Historically sockeye was canned, but more recently it has gained acceptance as a fresh or fresh-frozen product. In either case the importance of salmon farming is considerably more pronounced in these desirable species than in the context of total salmon supply. Also, the share of farmed salmon has increased quite substantially in the period under consideration.

The United States has the largest salmon fishery in the world (see Appendix 2, Table A4). Between 1981 and 1987 the US accounted for 44 per cent of all salmon taken in the capture fishery. Japan, despite having lost access to its traditional high seas fisheries through the international adoption of 200 mile fishing zones, accounted for 27 per cent of the commercial catch. Other important nations are the USSR (16 per cent) and Canada (12 per cent). The small amount *not* caught by the four leading harvesting nations is for the most part attributable to Atlantic salmon taken by North Atlantic based fishermen.

As is the case with the capture fishery, salmon aquaculture has been dominated by a few producing nations in recent years. Table 1.4 gives farm production data for the years 1981–1988.

Norway produced 61 per cent of all farmed salmon over the 1981–1988 period. In 1988 Norway had a market share of 59 per cent of world farmed salmon production and the three leading producers, Norway, Scotland and Japan, accounted for 81 per cent. However, most forecasts predict an erosion of market share for the current top producers, Norway included. Nonetheless, Norway will remain by far the largest supplier in the foreseeable future. Table 1.5 shows actual production for 1987–1988 and forecasts for the years 1990 and 2000.

Even if most forecasts prove high, it is clear that the next few years will be dynamic ones for the industry worldwide. It is also clear that the significance of farmed salmon to the world supply of the most desirable species will continue to increase. Table 1.6 provides a hypothetical production share for farmed salmon for the year 2000, assuming gains predicted in Table 1.5 are fully realized.

Table 1.4   Production of Farm Reared Salmon, by Country, 1981–1988 ('000 tonnes)

	Norway	UK	Japan	USA	Faroe Islands	Ireland	Canada	Chile	Others	Total
1981	8.4	1.1	1.1	0.5	0.1	Neg.	0.2	0.1	0.1	11.6
1982	10.7	2.2	2.1	0.7	0.1	0.1	0.3	0.1	0.2	16.5
1983	17.3	2.5	2.9	0.9	0.1	0.3	0.3	0.1	0.2	24.6
1984	21.9	3.9	4.4	1.2	0.1	0.4	0.3	0.1	0.3	32.6
1985	29.5	6.9	6.8	1.3	0.5	0.7	0.3	0.1	1.0	47.1
1986	44.8	10.3	7.2	1.4	1.4	1.3	1.0	1.2	1.4	70.0
1987	47.4	12.7	13.0	1.5	2.5	2.2	2.6	1.8	3.8	87.5
1988	80.3	22.0	16.6	3.8	3.1	4.7	9.7	3.5	5.5	149.2

Sources: See Section 1.3.

Table 1.5 Production of Farmed Salmon, by Country, 1987–1988, 1990 and 2000 (Forecast) ('000 tonnes)

Country	1987	1988	1990	2000
Norway	47.4	80.3	150.0	200.0
Scotland	12.7	18.0	40.0	70.0
Ireland	2.2	4.1	10.0	20.0
Faroe Islands	2.5	3.1	10.0	15.0
Iceland	0.5	1.0	2.0	5.0
Canada				
British Columbia	1.2	6.5	14.0	50.0
Eastern Canada	1.4	3.2	5.0	10.0
Chile	1.8	3.5	15.0	30.0
Japan	13.0	16.6	25.0	40.0
USA	1.5	3.8	8.0	10.0
New Zealand	1.3	1.5	3.0	5.0
Others	2.0	3.0	4.0	5.0
Total Production	87.5	144.6	286.0	460.0

Sources: See Section 1.3. Where production forecasts are not given in the text, they are the author's estimates.

Table 1.6 Production Share of Farmed Salmon in the Year 2000 (tonnes)

Wild catches [a]		
Chinook, coho, Atlantic	79 000	
Chinook, coho, Atlantic, sockeye	–	243 000
Farmed production (forecast)	460 000	460 000
Total production	539 000	703 000
Production share of farmed salmon (%)	85	65

[a] Wild catch figures are used for the following years and species: 1982 – chinook, coho; 1983 – sockeye; 1984 – Atlantic; see Table A1, Appendix 2.

Table 1.6 uses the highest annual landing totals for each species considered in the six year period 1980 – 1985. There are few reasons to believe that these totals will on average be exceeded with any regularity in the coming years. Indeed, given the current state of wild Atlantic and chinook stocks, production declines seem the more likely occurrence. To the extent that wild production of the 'glamour' species declines, salmon aquaculture's share will rise.

The figures in Table 1.6 clearly show that in a few years farmed salmon will dominate the markets for Atlantic, coho and chinook salmon. A production share of 85 per cent is predicted for the year 2000. Even if sockeye is added to the list of desirable species, farmed salmon will be the most important.

In 1988, Atlantic salmon accounted for about 80 per cent of total farmed production. Based on the production forecasts for 1990 and 2000, it can be seen that the Atlantic salmon's share of total production will decline in the years ahead as Canada, Chile and Japan emerge as major producers of Pacific salmon. And although Atlantic salmon are now being introduced in many Pacific rim nations, these experiments may be slow to develop and may well be insufficient to offset increased Pacific salmon production.

1.3 THE EMERGENCE OF SALMON AQUACULTURE

As the expansion in salmon supply will for the most part come from aquaculture, the following sections provide country by country sketches of established and emerging salmon farming industries. Special emphasis is placed on the development in Norway which has served as a model for many other countries.

Salmon aquaculture in Norway

Norway has long been the world leader in farmed salmon. Production figures for 1974–1989 are given in Table 1.7 along with farm and pen volume data.

Based upon the data in Table 1.7, it can be inferred that average farm size in terms of utilized capacity (the pen-volume actually used for rearing salmon) in 1987 was about 6 280 cubic metres. This was less than the government imposed licence limit, which was 8 000 m³ at the time. Observed farm size is a consequence of government regulations and the limited availability of inputs such as capital and until recently, smolts. Moreover, many farmers invest gradually as it takes three to five years to develop a new farm.

As of 1988, the licence limit was increased to 12 000 m³. However, each farm needs government authorization to increase its size to this limit. Such authorization depends on, among other things, environmental considerations. When fully developed, most farms can be expected to reach the imposed size limit. (A few farms, established prior to the introduction of licensing in 1973, are larger than this limit.) Similar technology is employed by most farms. Many have on-site processing facilities wherein fish are slaughtered, bled and gutted.

In 1989, 790 farms were in operation with a potential aggregate pen-volume of about 9.5 million m³. With an average production of 250 t per farm, production could reach 200 000 t as unexploited capacity is utilized. It is difficult to predict when these production figures will be realized. In the past, a lack of

Table 1.7 Norwegian Salmon Production 1974–1989

Year	Number of Fish Farms	Utilized Pen Volume ('000 m³)	Salmon Production (tonnes)
1974	189	1012	601
1976	168	928	1431
1978	219	1166	3540
1980	307	1581	4312
1981	316	1700	8418
1982	387	1999	10695
1983	411	2168	17298
1984	489	2440	21881
1985	562	3386	29473
1986	643	3959	44831
1987	747	4690	47417
1988	786	—	80250
1989 [a]	790	—	115000

[a] Preliminary.

Source: Norwegian Official Statistics (1987) for 1974–1986 and total production 1987; Fish Farmers Sales Organization for other data.

smolts has slowed down farm production, but large investments in smolt production have removed this bottleneck. In the future it is more likely that availability of capital will be the limiting factor. Production is expected to reach 150000 t in 1990. Total production capacity may be realized in the early 1990s.

A further expansion will result when the number of farm licences is increased. It is uncertain when new farm licenses will be issued. (There were also a few unused licences in 1988.)

Government financial assistance

The Norwegian government has played a very important role in financing the aquaculture industry through the Regional Development Fund (RDF). The RDF is responsible for funding new industrial activities in outlying areas. Total financial assistance for the period 1961–1987 was 1586.3 million Norwegian kroner (NOK) (Table 1.8).

Financial assistance is given in three ways, direct loans (at subsidized interest rates), investment grants of up to 40 per cent of capital investments (depending on location) and loan guarantees. The latter are particularly important to fish farming as financial institutions may be reluctant to accept fish stocks as collateral for loans.

Table 1.8 Government Financial Assistance to the Aquaculture Industry 1961–1987 (million NOK)[a]

	Loans	Investment grants	Loan guarantees	Sum
1961–1986	388.0	212.4	678.9	1 279.3
1987	66.0	59.0	182.0	307.0
Sum	454.0	271.4	860.9	1 586.3

[a] Nominal figures.

Sources: Fiskeridepartmentet (1986–87) for 1981–1986; Regional Development Fund for 1987.

Employment

A survey[1] undertaken in 1986 found that plants in the industry (broodstock farms, hatcheries, smolt operations and grow-out farms) average 4.4 employees. Indirect employment in secondary activities such as feed and equipment supplies, sales and distribution is thought to be roughly the same as the direct. In 1988 direct and indirect employment in the industry exceeded 10 000.

Government regulations[2]

A government licence is required to operate a fish farm in Norway. Licensing of fish farms was introduced as early as 1973. The law was, however, applied very liberally and all applicants were issued licences until 1977. Since then the government has controlled both the number of new licences issued and the size of farms. The present law regulating the industry dates from 1985. Grow-out farms are regulated as to:

(1) Entry.
(2) Location.
(3) Farm size (measured in pen volume).
(4) Ownership.

In addition to controlling farm size, the government controls the total number of licences issued, i.e. potential industry supply. The original rationale for this policy was to adapt production to market demand through a 'balanced development' of the industry. Recently, 'balanced development' has also been considered in relation to the capacity of the veterinary and extension services, education and research. This provides a further argument for a restrictive approach to awarding new licences. The Fish Farmers Association also favours 'balanced development', allegedly for the same reasons.

According to the licensing regulations, one firm (person) may hold a majority interest (51 per cent or higher equity share) in only one fish farm. Moreover, if a majority interest in a farm is to be transferred or sold, the transaction must be authorized by the Directorate of Fisheries. However, the extent of minority interests held by a firm (person) in fish farms is not regulated and is difficult to document.

The rationale for ownership regulations is that the government wanted an owner–operator structure in the industry. The result is that most farms are single-plant operations. The exceptions are companies with multi-plant operations established prior to licensing.

Since 1985, broodstock farms, hatcheries and smolt producers have been regulated by environmental criteria and size only. Smolt producers are permitted a maximum production capacity of 1 000 000 smolts.

As farm size is limited to 12 000 m^3, the regulations may prevent exploitation of economies of scale in production if efficient farm size is larger than the limit. Ownership regulations restrict horizontal integration and imply that most firms will be single-plant operations. As a consequence the industry will be prevented from exploiting economic benefits due to horizontal integration, e.g. common management and financing of farms, joint purchasing, joint harvesting and processing. Some (but not all) of these benefits can be achieved through cooperation between farms. It can be noted that several farms cooperate with respect to purchasing feed and processing.

The regulations limit firms' potential for investment in Norway. This is particularly true for large enterprises seeking further expansion. Consequently many Norwegian enterprises invest in fish farming abroad.

Barriers to entry

Barriers to entry exist when established enterprises can earn economic rents without attracting market entry. Economic rents are payments to factors of production above their opportunity costs. Norwegian producers have until recently been earning economic rents. Barriers to entry in the Norwegian industry exist in that the number of licences is restricted. There were about 2 500 applicants for the 150 licences awarded in 1985/86.

The physical and environmental characteristics of a farm site influence its productivity and the availability of infrastructure such as roads and electricity influences development costs (cf. Chapter 5). The long coastline of Norway, with its well-developed infrastructure, offers a large number of potential sites. Site availability has not limited industry expansion.

Smolt shortages have in the past limited expansion and prevented many farms from utilizing full capacity. In recent years high returns have attracted significant investment and this has removed the bottleneck for smolts. Prior to 1985 smolt producers were licensed according to the same strict criteria as fish farms. The result was an imbalance in the production of smolts and the capacity of the farming industry.

Smolt producers need to be located close to a stable source of fresh water, usually a river. The more liberal awarding of licences for smolt operations implies that the 'best' sites may already be developed. However, if the permitted maximum production of smolts (1 000 000 per unit) was increased, there would be a substantial potential for increased production.

The rents earned in fish farming have in the past meant few financial barriers to entry. Ownership of a licence has been sufficient to obtain financing. Moreover, the government has offered grants and subsidized loans to farms in fringe areas, although the amount of government support is expected to decrease. Compared to many other industries, the amount of investment required to achieve an efficient plant size is modest.

In summary, apart from the government's licensing regulations, there are no true barriers to entry in Norway.

Vertical integration

There are legal constraints on ownership of fish farms insofar as one firm (person) may hold majority interests in one farm only. This limits the ability of smolt producers to integrate forwards and purchasers to integrate backwards. While farmers may integrate forwards or backwards, few have the financial strength such investments require. This is particularly true for forward integration.

While the extent of vertical integration is limited, purchasers/exporters may have minority interests or purchase agreements with fish farms. This kind of vertical linkage is becoming increasingly common.

Distribution is controlled at the farm level. Trade in raw salmon is by law organized by the Fish Farmers Sales Organization which has the right to set minimum prices for salmon and to issue purchase licences. In 1989 there were 72 authorized purchasers.

To export salmon the exporter must meet the requirements of the Fish Export Act of 1955. In 1989, 72 firms exported salmon.[3] All these were also purchasers. In 1988, the 10 largest exporters accounted for 70 per cent of exports.

The industry regulations limit vertical integration and give rise to a fairly strict division between primary production and distribution (purchasing and exporting). From an economic point of view, this division seems unfounded. Fully integrated enterprises (ones that include all stages of the production process) would have better control of product quality than is the case today. Moreover, production and consumption would be brought closer together, which is important as products need to be adapted to market demand. Furthermore, there is a minimum size for an efficient sales and export organization. Permitting exporters to integrate backwards in order to ensure sufficient supplies could reduce fixed costs per unit sold. The present set of regulations prevents the industry from attaining some of the benefits that vertical integration offers.

Salmon aquaculture in Scotland

Salmon farming has existed as an industry in Scotland since the 1960s, but the major expansion has taken place in the 1980s. Fish farmers have raised trout in Scotland for more than 100 years. This experience has been beneficial to the salmon industry.

The basic conditions for farming in Scotland are similar to those found in Norway, but the coastline is less protected. This means fewer suitable sites for aquaculture and limits expansion with conventional sea-pen systems. The main areas for salmon farming are on the north and northwest coasts as well as on the islands. The east coast is too exposed to permit farming. In some areas, the fish tend to reach maturity at an earlier age than in Norway. This is a problem related to smolt quality.

The production of salmon in Scotland is given in Table 1.9. The 1987 production of 12 721 t was produced by 126 firms operating 185 sea-pen sites and

Table 1.9 Production of Salmon in Scotland 1980–1988

Year	Production (tonnes)
1980	598
1981	1 133
1982	2 152
1983	2 536
1984	3 901
1985	6 921
1986	10 300
1987	12 721
1988	17 951
1989[a]	33 000
1990[b]	40 000

[a] Preliminary figure.

[b] Forecast.

Source: Shaw and Rana (1987) for 1980–1986; Department of Agriculture and Fisheries for Scotland (1987) for 1987–1990.

11 land based sites.[4] Average weight per fish was 2.65 kg (this is less than in Norway). Fifty-seven per cent of the production in 1987 was fish harvested after only one year in sea water. (Salmon maturing after one year are called 'grilse'.) Smolt production and salmon farming directly employed 1144 people.

The Scottish industry is presently in a period of very rapid expansion. Total production is expected to reach 40 000 t in 1990 and 70 000 t by the year 2000. This 'ceiling' on output is essentially determined by site availability.

The United Kingdom domestic market absorbs about one half of the Scottish output. Having a substantial and stable home market is a definite advantage in terms of demand and transportation costs. Other important markets are the European Economic Community (EEC) and the USA. As an EEC member, Scotland does not meet the tariffs that face Norwegian exports to the EEC. The distance to the US market is about the same for the two countries.

Industry structure

While there is no licensing system as in Norway and no upper limit on farm size, starting a farm requires a lease from the Crown. As the best sites have been occupied, leases have become more difficult to obtain. A system to collect economic rents was introduced in 1987. The rate is £ 50 per tonne for farms with production greater than 50 t and £ 45 per tonne for farms with production less than 50 t.

As is the case in Norway, the government has actively encouraged aquaculture development. Financial assistance is given to small farms through the Highland and Islands Development Board (HIDB). In the period 1965–1987, the HIDB provided financial support worth £ 50.7 million to the industry. Most of this was in the form of grants; some was in loans. In the same period, the HIDB supported research with £ 4.1 million. A number of programmes supported by the EEC also provide funding.

The industry structure in Scotland is dominated by multinationals rather than small-scale farms. Many firms produce their own smolts. There is a high degree of vertical integration between hatcheries/smolt producers and grow-out farms. In general, the technology is more labour intensive than in Norway.

The size distribution of salmon producers in 1986 is given in Table 1.10. Three firms each produced more than 1000 t; combined they accounted for about 50 per cent of output. Despite the rapid increase in production, these three firms have maintained (or slightly increased) their share of output.

Marine Harvest, a subsidiary of Unilever, in 1988 had a production of 5500 t from 18 sites and is the largest salmon producer in the world. Its production forecast for 1990 is 10 000 tonnes.[5] The second largest is McConnel Salmon, a subsidiary of the trading group Booker McConnell, with 14 sites and a 1988 production of 1900 t, estimated to exceed 3000 t in 1990.[6] Third is Golden Sea

Table 1.10   Size Distribution of Salmon Producers 1986

Production tonnage of firms	Distribution of production		Distribution of firms (%)
	Quantity (tonnes)	Share of production (%)	
1–50	1 200	12	50
51–200	1 800	17	33
201–750	2 200	21	9
751–1000	–	–	–
over 1000	5 100	50	3
Total	10 300	100	95[a]

[a] Five per cent of the firms gave no response. These are all firms with low production.

Source: Shaw and Rana (1987).

Produce, a subsidiary of the Norwegian Norsk Hydro group.

This industry structure is very different from the Norwegian, where concentration at the production level is low. In Scotland, there is to some extent a dual industry structure. On one hand, there are the three large firms dominating the industry; on the other, there are small producers that were initially subsidized by the HIDB. However, a group of middle-sized producers is also emerging.

Salmon aquaculture in Japan

Japan is the world leader in farming Pacific salmon. In recent years the Japanese have raised more Pacific salmon (almost all coho) than the world's other producers combined. Atlantic and sockeye salmon are also reared on an experimental basis. Table 1.11 provides a historical record of farm production.

The majority of Japan's salmon farms are in Miyagi and Iwate prefectures on the island of Honshu, north and east of Tokyo. Japanese growing operations are small, typically with less than 1000 m^3 capacity. This is a result of government regulations requiring fish to be raised by individual small operators who are members of fishermen's cooperatives. The large fishing companies are not licensed for grow-out operations but they participate in all other aspects (smolt production, feed supply, sales brokerage, etc.) of the industry. Through arrangements with the growers, the large fishing companies (Nichiro, Taiyo, Nichimo) establish their own production groups and they control most of the production.

Eggs are imported for aquaculture operations. Japan has an extensive hatchery system for enhancing wild runs of chum salmon but does not produce

Table 1.11   Japanese Salmon Aquaculture Production 1978–1989

Year	Production (tonnes)
1978	80
1979	450
1980	1 500
1981	1 100
1982	2 100
1983	2 900
1984	4 400
1985	6 800
1986	7 200
1987	13 000
1988	16 600
1989[a]	21 000

[a] Preliminary figure.

Source: Atkinson (1987) for 1978–1985; Ministry of Agriculture, Forestry and Fisheries (Japan) for 1986; Atkinson (1989) for 1988–1989.

nearly enough coho eggs for salmon farming. They must be imported from the United States. The US is also a source for the small volumes of chinook eggs imported for rearing experiments the past few years. Attempts are being made by the Nichiro Fishery Company to culture chinook for the market.

Coho are put in sea pens in October through early December and are harvested the following May through July. The fish cannot be kept over the summer because of high water temperatures. The fish will weigh around 2 kg at harvest. Japan is unique among major growers of farmed salmon in that it consumes almost all of its own production.

Salmon aquaculture in British Columbia

Salmon farming began in British Columbia in 1972. The first farm was established on the Sechelt Peninsula, a region in southern coastal British Columbia northwest of Vancouver. This area was previously known to tourists and residents alike as an excellent place to catch wild salmon and the conditions that made it so must have seemed propitious to pioneer and subsequent fish farmers. By the mid 1980s nearly half of the aquaculture industry was located there. However, expansion of the industry was a slow process.

The basic supply conditions confronting the BC industry are similar to those found in Norway, with some important exceptions. The BC industry emphasizes the rearing of Pacific salmon species, particularly chinook and coho. There are

Table 1.12 British Columbia Salmon Aquaculture Production 1984–1988

Year	Coho (tonnes)	Chinook (tonnes)	Atlantic (tonnes)	Total production (tonnes)
1984	64[a]	43	–	107
1985	71[b]	50	–	121
1986	299	89	–	388
1987	700	500	–	1 200
1988	1 750	4 500	250	6 500

[a] Includes rainbow trout.

[b] Includes steelhead.

Source: Department of Fisheries and Oceans, Canada and BC Salmon Farmers Association.

several important biological differences between these species and the Atlantic salmon farmed in Norway. Atlantic salmon remain in fresh water considerably longer before smoltification than do Pacific salmon. Consequently Pacific salmon smolts are considerably cheaper to produce than their Atlantic counterparts. Atlantic salmon can tolerate greater densities in sea pens than can chinook or coho; thus, productivity per unit of pen volume is greater in Norway. The waters off the British Columbia coast tend to be warmer than those of Norway; this is generally considered a positive attribute. Finally, the infrastructure (roads, hydroelectric power, telecommunications, shipping links) is far more developed along the Norwegian coast. This has obvious impacts on the cost of developing and operating sea farms.

The province has a great potential for farm sites and there are no significant barriers to entry attributable to conventional sources such as capital costs.

Demand conditions confronting the two industries are basically the same. Both target the fresh market and both view North America as a major market for their outputs. Given the predicted increase in British Columbia production,

Table 1.13 Active Hatchery and Farm Sites in British Columbia[a]

	Hatcheries	Farm sites
May 1986	20	54
November 1987	37	118

[a] Some sites will have both hatchery and grow-out operations. Thus, the total number of active sites will be less than the sum of hatchery and farm sites.

Sources: DPA (1986) for 1986; British Columbia, Ministry of Agriculture and Fisheries (1988) for 1987.

these two producing areas will ultimately come into head-to-head competition in world markets for fresh, pen-raised salmon.

Production of farmed salmon in British Columbia is given in Table 1.12. Production increased sharply in 1987 and the expansion in coming years could be substantial. British Columbia is expected to become one of the world's major producers of farmed salmon. Total output in 1990 is forecast at 14000 t.

Through 1987 more coho than chinook were produced. The emphasis on chinook, as reflected in the 1988 figures, presumably reflects a market preference for larger fish. It can also be noted that Atlantic salmon have been introduced in British Columbia.

Industry structure

In 1984, there were only 10 active sites producing farmed salmon in British Columbia. The development since then is given in Table 1.13.

The figures in Table 1.13 are indicative of the great expansion in the British Columbia salmon aquaculture industry. In November 1987, there were 118 active farm sites. In addition, 35 farm licences were issued and 144 applications were being investigated. It is expected that about half the applications under investigation will be approved. This means that British Columbia could have 225 licensed farm sites in 1988. (As many applicants lack financing, not all of these will be active.) Assuming an output level of 300 t per site, the potential production from British Columbia could be close to 70000 t.

Average farm size in British Columbia (in terms of pen volume) is 27906 m³ (Bjørndal and Schwindt, 1987). This contrasts sharply with the 12000 m³ maximum imposed in Norway. However, this comparison is misleading. Stocking densities average between 20 and 25 kg/m³ in Norway as compared with planned maximum stocking densities of 7–8 kg/m³ in British Columbia. These low densities in BC are attributable to the nature of the species being reared.

Horizontal integration is common at the farm level. Although the pattern of vertical integration is unsettled, the large firms are commonly integrated from broodstock into grow-out and some even into processing. However, most farms have not integrated forward into processing. The traditional packing industry has substantial under-utilized capacity during the off-season for the commercial fishery.

A primary reason for the substantial expansion in British Columbia is the successful Norwegian development. The strict regulatory environment has led many firms to invest abroad. The Norwegian enterprises that have invested in British Columbia are large relative to the capital requirements in aquaculture. More importantly, these firms have exhausted investment opportunities in Norway in that regulations preclude them from entering or expanding in specific aquaculture activities. Opportunities may still exist in Norway, but they are not open to the firms in question.

A recent analysis reported that Norwegian interests were represented in roughly 50 per cent of all investments in the British Columbia industry (Bjørndal and Schwindt, 1987).

Salmon aquaculture in eastern Canada

Atlantic salmon are now being farmed in all of Canada's eastern provinces, although much of the activity is experimental or small in scale. The limiting factor is cold water and only New Brunswick, which utilizes the southern reaches of the Bay of Fundy, has achieved significant production levels. New Brunswick produced 1300 t in 1987 (an estimated 3000 t in 1988) while the rest of eastern Canada produced less than a 100 t. Thirty-two farms in New Brunswick, 10 in Nova Scotia and a land-based operation in Quebec accounted for nearly all of eastern Canada's production (Communications Directorate, DFO, 1988).

Throughout the Maritime provinces (Prince Edward Island, New Brunswick, Nova Scotia) and Newfoundland a number of salmon rearing experiments are on-going as methods are being sought to counter the limitations imposed by coastal waters that ice over or are lethally cold in winter. The region is home to a number of land-based trout farms and many of them are looking for ways to incorporate salmon aquaculture into their operations.

As with most emerging salmon farming industries, there has been a smolt shortage in eastern Canada. This will be a short run problem. New Brunswick produced approximately 820 000 smolts in 1987 and estimates were for 1 500 000 in 1988. A hatchery with a 200 000 smolt capacity has recently opened with partial production in Newfoundland, a province without significant salmon production at this time. In the long run there will be sufficient smolts for eastern Canada; the availability of sites, given winter water temperatures, will continue to be a problem.

Recently there have been small-scale experiments involving Pacific salmon in the province of Ontario. Attempts are being made to raise coho in the Great

Table 1.14 Salmon Production, Eastern Canada, 1984–1988 (tonnes)

1984	250
1985	400
1986	650
1987	1 400
1988	3 200

Source: Estimates derived from Communications Directorate, Department of Fisheries and Oceans (1988).

Lakes. It is far too early to predict the outcome of such experiments but it must be noted that a thriving coho sports fishery has been established in the region.

Salmon aquaculture in Chile

Salmon aquaculture in Chile is carried out in two modes; ranching, which dates from the early 1970s, and farming which began in 1979. Salmon is not a native species to Chile. Eggs are at present imported while local broodstocks are developed. Coho is the predominant species; chinook and Atlantic salmon are also being introduced. Because salmon are not native to Chile, its waters were originally free from viral salmon pathogens. Consequently, the industry has been relatively disease free so far, but this appears to be changing.

As Chile is located in the southern hemisphere, its seasons are opposite to those of the northern hemisphere. This provides a major advantage in competing for northern hemisphere fresh salmon markets. The off-season for the North American capture fishery (November to May) coincides with the harvesting season in Chile (January to March). Therefore, Chilean farmed salmon faces no competition from the capture fishery. Further, in the southern hemisphere maximum size is reached in winter, a season when there is little growth in the northern hemisphere. This is especially true for coho, which in Canada have a live weight of less than 1 kg in January while in Chile they weigh 2 kg when harvested after one year in sea water.

Most fish farms are located in the 10th region of Chile, about 1 000 km south of Santiago, near the city of Puerto Montt, and on Chileo Island. While there is a railway connection and a good road to Puerto Montt, the infrastructure south

Table 1.15   Chilean Production of Farmed Salmon, 1981–1990 (tonnes)

Year	Quantity
1981	70
1982	80
1983	94
1984	104
1985–1986	1 200
1986–1987	1 753
1987–1988	3 500
1988–1989[a]	5 500
1989–1990[b]	14 700[c]

[a] Preliminary   [b] Forecast

[c] Of this, 4000 t will be Atlantic salmon.

Source: Mendez (1988).

of Puerto Montt is quite poor. Although there is potential for expanding production in the Puerto Montt area in terms of site availability, the government is actively promoting expansion of the industry further south as part of its regional policy.

Table 1.15 provides estimates for Chilean farmed salmon production. The industry is expanding and Chile is expected to become a major world producer. In 1979, two farms were in operation; by March 1988 152 licences had been issued, with 45 farms in operation (Mendez, 1988). Seventy-one licences were issued for hatching or smolt production, 96 for grow-out and five for ocean ranching.[7]

Salmon aquaculture in Ireland

The Irish industry has marked similarities to that of Scotland; both produce Atlantic salmon, the majority of production comes from a few large farms, there is considerable foreign investment and good sites are at a premium. Site availability is the most important constraint on expansion. All sheltered sites are already occupied. This has forced the industry to experiment with cages that are more weather-resistant than ordinary sea-pen structures.

Table 1.16 gives recent production figures. Expansion has been less rapid than in Scotland and future potential is also less due to limited site availability. Production for 1990 is forecast at 10 000 t. As in Scotland, there is a high proportion of fish maturing after one year (grilse).

Growers in Ireland have suffered reversals. Hot summers (1984 in particular) conducive to disease and ocean temperature problems, along with wave exposure damage, have kept production growth in check. Also, a shortage of

Table 1.16 Salmon Aquaculture Production in Ireland 1981–1988

Year	Production (tonnes)
1981	35
1982	103
1983	257
1984	385
1985	722
1986	1215
1987	2232
1988	4075
1989	6200 (prelim.)

Sources: Irish Sea Fisheries Board for 1981–7, Irish Aquaculture Association for 1988–9.

smolts has mitigated against large production increases; eggs have been imported from Norway throughout the 1980s to help meet the shortfall. Now the smolt problem is being alleviated by large investments in hatchery and smolt rearing facilities by salmon growers (often large multinationals). The Electricity Supply Board (ESB) has long had hatchery programmes in place and operates hatcheries and grow-out farms.

Farmed salmon is for the most part exported to the United Kingdom and France. In 1985 the three leading producers accounted for 65 per cent of all production.

Salmon aquaculture in the Faroe Islands

Throughout the 1950s and 1960s efforts were made to launch salmon aquaculture in the Faroes. They did not pay off. In 1973 the government began to experiment on two state-owned farms and by the early 1980s private farms were in operation. Table 1.17 gives historical production figures. Production increased rapidly in 1986 and is expected to exceed 10 000 t in 1990.

In the early years smolt production lagged behind demand in the Faroes. Most smolts have been provided by a state-run hatchery and the importation of smolts is not permitted. A 400 per cent increase in smolt production between 1982 and 1984 eased the shortage to some extent.

The fjords of the Faroes offer well sheltered growing sites but in limited number. A licence is required to operate a farm. In 1988, 68 farms were in operation with a total *licensed* capacity of 720 000 m³. Of this about 74 per cent was in use (Reinert, 1987). Maximum size per farm is 15 000 m³ (Reinert, 1987). Foreign ownership is precluded. Water temperatures are favourable, although somewhat colder than Norway.

Table 1.17 Faroese Salmon Aquaculture Production 1982–1988

Year	Production (tonnes)
1982	60
1983	105
1984	116
1985	470
1986	1 370
1987	2 500
1988 [a]	3 100

[a] Estimate.

Source: Reinert (1987).

Salmon aquaculture in Iceland

Iceland is unique among salmon farming nations in that it has not been constrained by a smolt shortage, unlike all other emerging producers. This is due in part to a long history of stocking Icelandic waters with the produce of small hatcheries. Further, the government established an experimental fish farm in the early 1960s, in large part to work on smolt development.

Ocean ranching of salmon has long been carried out in Iceland. Return rates ranged from 3–14 per cent in the period 1982–1987. The catch from ocean ranching was 40 t in 1987 (Jóhansson, 1988). However, it is salmon farming that has the most growth potential. It is estimated that Iceland in 1988 had in place the capacity to raise over 6000 t of salmon (Jóhansson, 1988). Further, it can produce the smolts necessary to do so. The chief constraints are low water temperatures and a lack of sheltered locations. The response to these conditions is on-shore operations which are more common in Iceland than elsewhere and are more expensive than sea-pen operations.

Table 1.18 Salmon Aquaculture Production in Iceland 1986–1987

Year	Production (tonnes)
1986	123
1987	490

Source: Helgason (1987), Jóhansson (1988).

In 1987 the production from sea-pen and land-based farms was 223 and 267 t respectively (Jóhansson, 1988). Clearly the growth of on-shore farming will be directly tied to the price farmed salmon is able to command. In a world market that is experiencing a dramatic increase in supply, the cost of on-shore farming may well be a constraint on Iceland's ultimate production figures.

Salmon aquaculture in the United States

Commercial salmon farming has a relatively long history on the US west coast. Throughout the 1960s and 1970s coho salmon were raised to pan-size (about 0.5 kg) for restaurant and supermarket sales. In the early 1980s the move to producing large salmon began, although production of pan-sized fish is still the rule. There were several large-scale attempts at sea ranching during this period as well. They have been abandoned in subsequent years.

Although the US west coast began producing Pacific salmon (coho) there have been attempts to introduce Atlantic salmon. The oldest and largest farm in Washington state was sold (by the Campbell Soup Company) to Norwegian investors in early 1988. The new owners will shift production emphasis to Atlantic salmon, hoping to produce about 2000 t a year.

The unknown element in future US production is Alaska. Currently it is illegal to raise farmed salmon in the state. The moratorium on production was upheld by the Alaska State Legislature in the summer of 1988 (Bjørndal, Schwindt and Marshall, 1989). A study prepared for the Alaska State Legislature (Pierce, 1987) estimates possible Alaskan production at 20 000 t by the year 2000.

On the east coast the state of Maine has begun to produce Atlantic salmon in small quantities. A number of experimental research programmes, designed to increase production are in place in hopes of establishing a viable industry in Maine.

Production figures for the United States are often questionable. There are reports of small coho production having reached 11 000 t in the mid 1970s (BIM, Dublin, 1986). This is difficult to verify. Table 1.19 gives estimates of US production from 1985–1988.

Table 1.19 United States Production of Farmed Salmon 1985–1988

Year	Production (tonnes)
1985	1 000
1986	1 500
1987	1 500
1988[a]	3 800

[a] Preliminary figure.

Source: Food and Agriculture Organisation of the United Nations, Rome, FAO/GLOBEFISH Highlights 4/88.

Regardless of the accuracy of past production statistics, the US is not likely to emerge as a major producer unless Alaska comes on line as a supplier of farmed salmon. The Washington and Oregon coasts offer only a limited number of good sites, and development even now is often a source of environmental conflict in those states. If the conflict between the capture fishery participants and would-be fish farmers in Alaska is resolved in favour of fish farming, the US could become a major producer.

Table 1.20   New Zealand Salmon Aquaculture Production 1984–1988

Year	Production (tonnes)
1984–1985	200
1985–1986	–
1986–1987	712
1987	1 300
1988[a]	1 500

[a] Preliminary figure.

Source: Todd (1987) for 1984–1987; Food and Agriculture Organisation of the United Nations, Rome, FAO/GLOBEFISH Highlights 4/88 for 1988.

Salmon aquaculture in New Zealand

Currently most farmed salmon in New Zealand is produced by two companies with sea cages in Big Glory Bay on Stewart Island.[8] Total production for 1988 is estimated to be 1500 t of chinook salmon. It is illegal to raise coho in New Zealand. New Zealand is the only producer in the world with a farm stock composed entirely of chinook. The fish are harvested after two years with harvesting taking place between October and March. Japan and the United States are the principal export markets. Table 1.20 gives New Zealand's production for the years 1984–1988.

Smolts in New Zealand have traditionally been supplied by a government-run hatchery to fish farmers who enjoy the same advantages as their Chilean counterparts; growing sites that are a long way from the natural range of Pacific salmon (and thus, their diseases) and coastal waters that are largely unpolluted. As a further means of disease control, the goverment very strictly regulates the importation of salmon products.

NOTES

1 Source: Arbeidsdirektoratet and Fiskeridirektoratet (1986).
2 See Bjørndal and Salvanes (1987) for a further analysis of the effects of government regulations.
3 Source: Fish Farmers Sales Organization.
4 Source for factual information in this paragraph: Department of Agriculture and Fisheries for Scotland, 1987.
5 Source: *Fish Farming International*, June 1988. The production figure for 1988 is preliminary.

6 Source: *Fish Farming International*, February 1988. The production figure for 1988 is preliminary.
7 The total number of licensed sites is 152. However, some sites are licensed for both hatchery and grow-out operations.
8 See *Seafood Business*, January/February, 1988.

2   Markets for salmon

Salmon is marketed as a fresh, frozen, smoked, salted or canned product. To some extent the different products utilize different salmon species. However, all products have a high quality image with consumers in their respective markets. In this chapter, consumption of salmon is described and future competition analysed.

2.1 SALMON CONSUMPTION

The major markets for salmon are found in Japan, the United States and western Europe. There are important differences between the markets with respect to species and product forms. To a large extent these differences can be explained by differences on the supply side and in consumer taste.

Throughout the western world there is increased awareness of the health aspects of food. In the past decade there has been a shift in demand towards 'healthier' sources of food at the expense of, for example, red meat. This change in consumption patterns has also benefited fish producers in general and salmon producers in particular. Increases in levels of personal disposable income have also led to greater demand for salmon.

The main producing nations of wild and farmed salmon (Sections 1.2 and 1.3) only partly overlap with the main consuming nations. Therefore, there is substantial international trade in salmon and salmon products. Domestic consumption in a country is equal to local production plus imports minus exports, adjusted for changes in the level of inventories.

Per capita consumption data for salmon and all other fish for selected countries are provided in Table 2.1. Japan has the world's highest per capita consumption of fish. It also consumes more salmon than any other country both in absolute and per capita terms. Other important markets for salmon include the United States, the United Kingdom and France.

Total consumption of salmon in Japan amounted to 313 000 t in 1987. Per capita consumption increased from 1.5 kg in 1980 (Ruckes, 1987) to 2.6 kg in

Table 2.1 Per Capita Consumption of Fish, by Country

Country	Per capita consumption of all fish[a] (kg)	Per capita consumption of salmon (kg)[b] 1987
Japan	69.3	2.6
Norway	41.2	1.7
France	25.8	0.7
Canada	22.4	0.3
United Kingdom	18.8	1.0
United States	18.5	0.5
World Average	12.4	0.16

[a] Average for 1984–1986, live weight in kilograms. Source: FAO Fisheries Statistics (1987).

[b] Atkinson (1989) for Japan; author's estimates for other countries. Inventories are assumed constant.

1987. Correcting for population growth, this represents almost a doubling of total consumption. No other market has exhibited such remarkable growth in this time period.

The main product in Japan is salted salmon, but salmon is now accepted as sashimi. The demand for sashimi and salmon roe is increasing, while the demand for salted products is declining.

With a total market quantity of about 130 000 t and a per capita consumption of 0.5 kg in 1987, the United States is the second largest market for salmon. Important product forms are fresh, frozen and canned. While the demand for canned products is declining, consumption of fresh and frozen products is increasing.[1]

France and the United Kingdom are also important salmon markets. In France about 70 per cent is consumed as a smoked product (Shaw, 1989), especially during the Christmas season. Salmon is imported raw (fresh/frozen) and processed locally. In 1987, consumption of smoked salmon in France was about 30 000 t (live weight; Shaw, 1989). The remaining quantity is consumed mainly as a fresh product. In the last couple of years consumption of fresh salmon has increased considerably, in the restaurant and the retail market.

In the UK salmon has traditionally been consumed as a canned product imported from North America. In addition, imported fresh/frozen Pacific salmon was used for smoking, while wild-caught Atlantic salmon was a very expensive fresh product. With the expansion in salmon farming in Scotland, the UK salmon markets have changed considerably. In particular, fresh consumption has increased.

Japan is by far the world's leading importer of salmon. In 1987 Japan imported almost 111 000 t of salmon, not including canned or salted roe

Table 2.2 Imports of Salmon (Fresh and Frozen) by Country 1987

Country	Imports[a] ('000 tonnes)	Principal suppliers by percentage[b]
Japan	111.0	US – 85%, Canada – 8%
France	39.0	US – 32%, Norway – 31%, Canada – 15%
United States	19.0	Norway – 41%, Canada 41%
Denmark	13.0	Norway – 59%, Canada – 14%
West Germany	11.5	Norway – 44%, Denmark 30%, Canada – 10%

[a] Sources: Shaw (1989) for West Germany; Food and Agriculture Organisation of the United Nations, Rome, FAO/GLOBEFISH Highlights 4/88 for others.

[b] Based on Shaw (1989), FAO/Globefish Highlights 4/88 and Norwegian export data.

products. Table 2.2 lists imports for 1987 for the major importing countries as well as principal suppliers.

The US currently supplies most of Japan's imports, the biggest part of which is wild sockeye (mostly frozen) which the Japanese salt in their traditional manner. The US is a major supplier in Europe as well. The overwhelming majority of US exports to Europe is ocean-caught, frozen or canned Pacific salmon. The frozen salmon is to a large extent used for smoking. In recent years, Norwegian and Scottish exports seem to be replacing US products in countries such as West Germany and Denmark. Exports of Pacific salmon to France have been quite stable in terms of quantity, but the market share has declined due to increased quantities of Atlantic salmon (Shaw, 1989).

Norway, the other major supplier to Europe, ships fresh farmed Atlantic salmon along with smaller quantities of smoked and frozen fish. In the 1980s Norway has become the world's leading exporter of fresh salmon with high market shares in the US and Western Europe.

In Table 2.2, Denmark appears as a substantial importer of salmon but also as a major supplier of salmon to West Germany. A substantial part of exports to Denmark is re-exported to other countries in the European Economic Community, largely in processed form (smoked).

Ocean-caught salmon are utilized very differently from farm-raised salmon. The wild catch of pink is used extensively for canning. Higher quality sockeye are mainly exported to Japan, while lower quality sockeye are used for canning. The predominant use of the Japanese chum catch is for salting. In addition, these three species are used as fresh, frozen products. Wild-caught coho, chinook and Atlantic salmon are used as fresh/frozen products, but also for smoking. Farmed salmon are predominantly marketed fresh but may also be used for smoking. Table 2.3 provides estimates for imports of farmed salmon. The US and France are by far the largest markets; in 1986 they accounted for 53 per cent

Table 2.3 Estimates of Farmed Salmon Imports (Fresh, Frozen, Smoked), by Country, 1986

Country	Farmed imports ('000 tonnes)	Fresh salmon as a percentage of farmed imports
United States	12.7	92
France	12.4	80
Denmark	6.2	82
West Germany	5.2	88
Sweden	2.0	85
United Kingdom[a]	2.0	80
Spain	1.9	95
Others	4.8	86
Total	47.2	85

[a] The UK figure is misleading as an indicator of consumption of farmed salmon due to the domestic (Scottish) production.

Sources: Shaw and Rana (1987); BIM (1986); *Seafood Business* (various issues); *Fish Farming International* (various issues).

Table 2.4 Estimates of Farmed Salmon as a Percentage of Fresh or Fresh/Frozen Consumption, by Country, 1985

	Farmed as a percentage of fresh	Farmed as a percentage of fresh/frozen
Spain	95	90
Denmark	90	65
France	90	50
West Germany	88	55

Sources: Shaw and Rana (1987); BIM (1986); *Seafood Business* (various issues).

of imports. About *21 000* t were consumed by the countries that produced them, notably Japan, Norway and the UK. Fresh salmon made up 85 per cent of farmed salmon imports.

Table 2.4 gives estimates for consumption shares for several European countries. The table shows that even when fresh and frozen salmon are lumped together as one product type, farmed salmon account for a large part of that market. When fresh sales alone are considered, cultured salmon dominates. The higher relative importance of farmed salmon in these countries, compared to the US, is due to smaller supplies of wild salmon in Europe.

Table 2.5 Norwegian Exports of Salmon to Selected Markets in 1981 and 1988 ('000 tonnes)

	France	USA	Denmark[a]	West Germany
1981	2.7	0.02	1.2	1.6
1988	18.7	9.9	14.2	7.5

[a] A substantial part of the exports to Denmark is re-exported to other countries in the European Economic Community, largely in processed form (smoked).

Sources: Bjørndal (1987) for 1981; the Export Council for Fresh Fish, Aalesund, Norway for 1988.

2.2 FUTURE COMPETITION

By 1990 the production of farmed salmon may be double that of 1988 (see Table 1.5). Are there markets that can absorb such quantities? To a certain extent one can sell more by entering new markets. Norwegian exports to the United States, France, Denmark and West Germany are an example of this. Table 2.5 demonstrates the kind of growth that is possible by opening and expanding markets in concert with strong increases in production.

The substantial increase in exports to the United States is especially remarkable. The American market has in the period of a few years developed into one of the most important markets for Norwegian farmed salmon. However, Table 2.5 also illustrates another aspect relevant to the continuing increase in world production. Farmed salmon has already established a presence in the most important geographical markets – Western Europe, the United States as well as Japan – and can only enter new market segments, e.g. salmon smokers, restaurants and supermarkets, in those countries. Exporting larger quantities to the already established markets will require better market coverage and deepening market penetration.

In the main consumer markets, farmed salmon competes with wild salmon,

Table 2.6 Per Capita Beef, Chicken and Fish Consumption, USA, 1976–1985 (kg per year)

Year	Beef	Chicken	Fish
1976	58.4	23.6	7.1
1980	47.3	27.6	7.2
1985	48.8	31.7	7.1

Source: Food Consumption Statistics, Organisation of Economic Co-operation and Development (OECD), Paris 1976–1985.

other seafood and food products. While wild salmon is clearly a substitute for farmed salmon, it is available fresh only in the harvesting season. During the rest of the year it competes only in the market for frozen products. The day-in and day-out competition for fresh farmed salmon comes from other seafood and other protein sources. Table 2.6 details per capita beef, chicken and fish consumption in the United States.

Clearly, North American consumers eat considerably more beef and chicken than fish. Certainly a part of the increase in chicken consumption shown in Table 2.6 is a result of a move towards 'healthier' sources of protein and away from red meat. The 'health' advantages that accrue to chicken over animal sources of protein are also present in fish. Market gains are available to salmon through direct competition with other protein sources.

Increased demand for salmon can come about through a change in consumer preferences. Throughout the western world there is a trend to increased consumption of low cholesterol foods such as poultry (Table 2.6). This development will likely also affect seafood products. As a traditional 'luxury' product, salmon consumption will also increase as personal income rises. Consumer preferences may be influenced by product promotion such as the one undertaken by farmers' associations in different countries. The opening up of new market segments will also increase demand. All these factors will combine to counter a downward pressure on prices caused by an increase in supply. However, production increases forecast in coming years may be so substantial that there will still be downward pressure on prices.

How far the price of salmon will decline depends on the elasticity of demand. The results of some recent demand studies for Atlantic salmon are summarized in Table 2.7. The demand for salmon is found to be elastic in both the United States and France, i.e. a given proportional decline in price will cause a proportionally greater increase in quantity demanded. Further, the demand for salmon is found to be highly income elastic.

In Table 2.7, only some of the most recent demand analyses have been cited. Earlier demand analyses are based on other species (coho, chinook) or product forms (frozen, canned) and data a few years old. As salmon markets have experienced significant changes in the last five years, these studies may tell little about current and future markets. A problem with the more recent studies is that they are based on fairly short time series. Accordingly, the number of data points is limited. Moreover, definition of a substitute product for fresh farmed salmon has proved difficult. Commonly, chinook has been chosen as the substitute product, but this is available in fresh form only during the harvesting season. All demand analyses presented in Table 2.7 are based on data at the exporter–importer level, i.e. they represent derived demand for salmon. Little, if anything, is known about demand elasticities at the point of final consumption.

Some of the elasticities presented in Table 2.7 are long run. Short run elasticities may be lower than long run. Therefore, a given increase in quantity

Table 2.7 Estimated Demand and Income Elasticities for Atlantic Salmon[a]

Geographical market	Time period	Own price elasticity	Cross price elasticity[b]	Income elasticity	Author(s)
United States[c]	Jan. 1983– March 1987	–2.51	0.35	5.56	Hermann and Lin (1988)
European Economic Community[c]	Jan. 1983– March 1986	–2.82	0.56	3.66	Hermann and Lin (1988)
United States[de]	Jan. 1985– May. 1987	–3.60	–	–	Singh (1988)
United States[e]	Jan. 1983– Dec. 1988	–2.00	0.95	2.11	DeVoretz and Salvanes (1990)
European Economic Community[e]	Jan. 1983– Dec. 1988	–2.14	1.12	2.14	DeVoretz and Salvanes (1990)

[a] All analyses are for Norwegian salmon.

[b] Whole chinook was specified as the substitute good.

[c] Long run elasticities are presented.

[d] In the Singh model specification, only the own price elasticity was estimated.

[e] Short run estimates.

may cause a greater price reduction in the short run than in the long run. Similarly, a rise in income will cause less of an increase in demand in the short run than in the long run.

As Shaw and Muir (1987) point out, the elasticity of demand will also vary over the lifecycle of the product. This is particularly relevant for salmon, as markets presently are undergoing substantial changes.

As output of salmon is increased, there will be a downward pressure on price and profit margins. The competition between the producers will become more intense and the price is likely to fall below cost of production for some producers, at least in the short run. The ensuing period of adjustment and instability in salmon markets may well be quite protracted due to the long production cycle for salmon (see Shaw and Muir, 1987).

Farm-raised salmon is to a large extent a bulk product with little or no product differentiation. The salmon are bled, gutted and exported fresh and unprocessed. In other words, farmed salmon appears to be a homogenous product. This is the underlying assumption for the demand analyses cited above. However, it may be argued that fresh salmon qualifies as a search good in that its product characteristics (e.g. colour, species, size, freshness, and general condition) are observable prior to purchase.

Some authors, notably Anderson (1987), stress that salmon is really a heterogenous product. The demand is a function of product attributes such as species, country-of-origin, production method, quality, fresh or frozen, and availability. In a survey of New England restaurants, fishmongers and restaurants, Anderson found that freshness, product form (whole or portioned), delivery time, price, minimum order quantity and species were the most important factors in determining demand. Factors such as colour and country-of-origin were found to be less important. The concept of a search or heterogeneous good assumes that buyers are knowledgeable about the different attributes of the various products. At the wholesale level, this is clearly the case. However, it is less likely that final consumers are able to identify and distinguish between different product characteristics. This is even more the case when the product is processed.

Most salmon is exported fresh. This requires fast, reliable transportation. Salmon is shipped by freezer or cooler trailer in chilled or refrigerated containers on the European and North American continents or by air to overseas markets. The question is whether the *distribution channels* will be able to absorb a substantial increase in quantity. Limited air-cargo capacity may become a constraint on shipments to North America both from Europe and Chile. For this reason, new products may have to be developed that go through other channels or new distribution systems may be developed. An example of the latter is high-speed 'well-boats' that now deliver live Norwegian salmon to ports on the European continent and Great Britain. As production quantity is increased, it is likely that more salmon will be traded in frozen form. This will permit shipment by boat to overseas markets.

Comparative advantages

With the exception of Japan and Scotland, all producers of farmed salmon are geared toward export markets. The production and distribution of farmed salmon is illustrated in Figure 2.1. This figure may be used to illustrate competitive advantages.

The various producer countries have different cost structures, depending on supply conditions such as site availability, production characteristics and technology, availability of inputs and regulations. As production and consumption generally take place in different countries, the products are transported, often over considerable distances. As the products are mainly exported in fresh form, efficient distribution systems are required both in the producing and consuming countries. Tariffs, trade barriers and exchange rate conditions differ

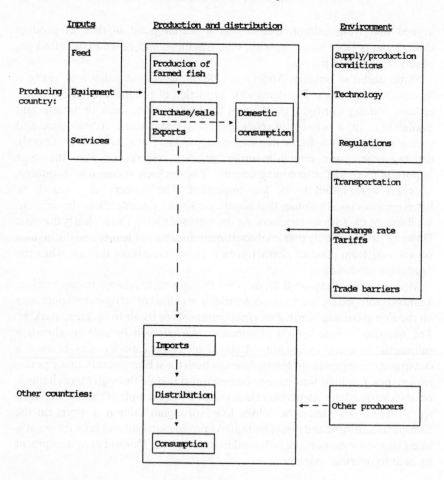

Fig. 2.1 Production and Distribution of Farmed Salmon

as well. These are all factors that influence competitiveness.

Tariffs can seriously affect competition. Scotland and Ireland are members of the European Economic Community (EEC) and can hence export to EEC countries without paying the tariffs that face non-EEC exporters. This is especially important for processed products which face the highest tariffs. (A two per cent import duty is levied on unprocessed salmon exported to the EEC by non-members, while the duty is 13 per cent for processed salmon. As of January 1989, the duty on fresh products is suspended.) These advantages become increasingly important as prices and profit margins fall.

A free trade agreement between Canada and the United States was implemented in January 1989. This involves free access for Canadian farmed salmon to the important US market. Although there are at present no tariffs on salmon imported into the US, it can be noted that if any protectionist measures were to be introduced in the US, Canadian producers would be exempted due to the free trade agreement and thus have a competitive advantage over other suppliers.

In the different markets, farmed salmon from different countries compete. For example, some of the producers (Norway, Scotland, Ireland and Canada) are better located in relation to important consumer markets (Europe and the US) than others are. The major market for Chilean farmed salmon is the United States during the off-season for the Pacific salmon fishery. Being able to supply fresh large coho at this time is a competitive advantage for the Chileans. On the other hand, the distance to the US market is 6000 km or more. This is equivalent to the distance from Europe to Los Angeles. Long transportation distances are a major disadvantage for Chilean and European producers delivering to the US markets when compared to producers in British Columbia. For European suppliers, this is particularly true for deliveries to the western US. On the other hand, prices of several inputs – labour, lumber and possibly feed – may be lower in Chile than in northern countries and cost of production may be less.

Due to transportation and distribution costs, the low-cost producer may not be the low-cost supplier in any given market. Chile appears to produce fish more cheaply than does British Columbia, yet it is probably not the low cost supplier in for example the US market. British Columbia's locational advantage is too great.

As pointed out (Section 1.1), the fish farmer may influence product attributes and develop products according to consumer preferences. Moreover, as preferences vary between segments, the farmer may develop specific products for each target group.

As yet marketing does not depend heavily on advertising. Branding is not very important and most promotion is done on a country-of-origin basis. It is also too early to tell whether producers will be successful in developing brand names.

NOTES

1 In the US, 75 per cent of imported salmon goes to catering (*Aquaculture Ireland*, January/February 1988, p. 34).

3 The theory of optimal harvesting of farmed fish – an introduction

The optimal harvesting time for farmed fish is considered in this chapter. Certain qualitative results regarding optimal harvesting are derived and analysed as to how they are influenced by different factors. A simple biological model, the basis for this analysis, is outlined. Thereafter different types of costs are introduced and optimal harvesting for each case is analysed. Examples of estimations of optimal harvesting for salmon and turbot are given.

The analysis is simplified in several ways. It is assumed that all plant investments are already undertaken and hence irrelevant to the decision process. Only the variable costs are relevant. Stochastic fluctuations in growth and uncertainty concerning parameter values are not considered. The price of fish is assumed to be fixed. These assumptions may seem to be very restrictive; they will be returned to later.

The analysis is carried out in a *continuous time* framework. This is correct in a theoretical analysis such as this one, as growth is a continuous process through time. This approach also constitutes the basis for harvesting models for actual fish farms. However, in such cases one would employ a *discrete time* model with the parameters updated on a regular basis. In the next chapter an example of a discrete time harvesting model for salmon will be given and the connection between these two types of models will be shown.

This chapter requires a certain basic knowledge of mathematics. Readers who do not possess such knowledge may still enjoy the verbal discussion. (Also refer to Chapter 4, which to a certain extent covers the same topic.) Only the simplest cases are included in this chapter. More advanced analysis is contained in Appendix 1, where some mathematical derivations are also shown.

3.1 A BIOLOGICAL MODEL

The production process in fish farming was discussed in Section 1.1. Here, mathematical equations are employed to illustrate the process. It begins when fish are released in an enclosure such as a fish farm or a fiord. The fish are called a *yearclass*, as all fish are of the same age. The number of fish released is for now considered a given parameter. Over time two biological processes will affect this yearclass: some individuals will grow and some will die. This biological process can be described by an adapted Beverton–Holt model for a single yearclass:

$$N(0) = R \tag{1}$$

$$\frac{dN}{dt} \equiv N'(t) = -M(t)N(t), \qquad 0 \leqslant t \leqslant T, \tag{2}$$

$$N(t) = RE^{-\int_0^t M(u)du}. \tag{3}$$

In this model, the variable t measures time from the release of fish as well as the age of fish. $N(t)$ is the number of fish at time t with dN/dt denoting its time rate of change. At the outset, $t = 0$, R fish – 'recruits' – are released (Equation 1). Over time the number of fish in a yearclass changes due to natural mortality. This is given by the instantaneous mortality rate $M(t)$, which can vary over time (Equation 2). $N(t)$ in Equation (3) is the number of remaining fish at time t.[1]. The fish live until age T. One can alternatively consider this the time of sexual maturity.

If the mortality rate is assumed to be constant over time, i.e.

$$M(t) = M = \text{constant},$$

equations (2) and (3) simplify to the following:

$$\frac{dN}{dt} \equiv N'(t) = -MN(t), \qquad 0 \leqslant t \leqslant T, \tag{2'}$$

$$N(t) = Re^{-Mt}. \tag{3'}$$

The development of the number of fish in a yearclass is illustrated in Figures 3.1(a) and (b). In Figure 3.1(a) the mortality rate is constant over time (see Equations (2') and (3')), but it varies in Figure 3.1(b) (see Equations 2 and 3).

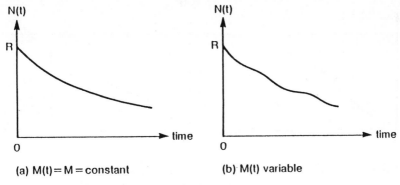

(a) M(t)=M=constant   (b) M(t) variable

Fig. 3.1 (a) and (b) The development of $N(t)$ over time

There is reason to believe that the mortality rates for both wild and farmed fish vary over time. For instance, many fish die when they are released in a fish farm.

Now let $w(t)$ symbolize the weight per fish at time t. The time rate of change in weight, i.e. the growth, is then $w'(t) \equiv \dfrac{dw}{dt}$. In general a *growth function* is expressed as follows:

$$w'(t) = g(w(t), N(t), F(t)).$$

Growth is here expressed as a function of three variables: *weight*, *number of fish* (density) and *feed quantity* ($F(t)$). The derivatives of the growth function with respect to the variables will explain how each variable influences growth.

In fish farming it is believed that excessive density can reduce growth, i.e.

$$g_N < 0.$$

If, however, density does not influence growth,

$$g_N = 0.$$

The relationship between feeding and growth is an interesting question. In traditional fisheries the fish find their nutrition in the sea. In fish farms they are fed. The condition for feeding to be undertaken is that

$$g_F > 0,$$

i.e. feeding actually increases growth.

The consequences of growth being dependent on density and feeding will be considered later. For now growth is expressed as a function of time, i.e.

$$w'(t) = g(t).$$

This function presupposes a certain density and feeding path. The weight of the individual fish at time t, $w(t)$, is then given by

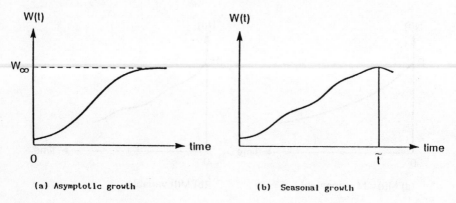

(a) Asymptotic growth   (b) Seasonal growth

Fig. 3.2 (a) and (b) Weight curves

$$w(t) = w(0) + \int_0^t w'(u)du.$$

$w(t)$ is equal to the weight when the fish is released, $w(0)$, plus growth until time t.

An example of a weight curve is given in Figure 3.2(a). In the example, fish grow towards an asymptotic value. In fish farming one can imagine a more marked seasonal growth. An example of this is shown in Figure 3.2(b). Seasonal growth is also common for many types of wild fish. Specifications of weight functions will be given in the empirical section below.

The individual fish will grow towards a maximum value either asymptotically or not (see Figures 3.2(a) and (b)). Define

$$w'(\tilde{t}) = 0$$

where $\tilde{t}$ is the time the fish reaches its maximum individual weight.

The biomass weight for the yearclass at time t, $B(t)$, is defined as

$$B(t) = N(t)w(t) = Re^{-Mt}w(t). \tag{4}$$

Here all fish are assumed to be of equal weight. Accordingly, the analysis is in terms of the 'representative' or the average fish. This assumption will be modified later.

Biomass weight over time is illustrated in Figures 3.3(a) and (b). $t = t_0$ is the time of maximum biomass weight. Figure 3.3(a) shows the normal biomass weight curve in the Beverton–Holt model, while Figure 3.3(b) presumably better illustrates a yearclass of farmed fish. The variations in yearclass weight can be a result of seasonal growth or variations in the mortality rate over time, or both. This will be dealt with later.

As with individual fish, the yearclass will also reach a maximum value. Changes in biomass weight over time are given by

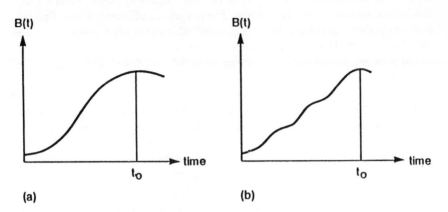

Fig. 3.3 (a) and (b) Biomass weight $B(t) = N(t)w(t)$

$$
\begin{aligned}
B'(t) &= w'(t)N(t) + w(t)N'(t) \\
&= w'(t)N(t) - Mw(t)N(t) \\
&= \left[\frac{w'(t)}{w(t)} - M\right] B(t).
\end{aligned}
$$

The derivations have made use of Equations (2') and (3'). $\frac{w'(t)}{w(t)}$ is the relative growth rate of the fish which presumedly *decreases over time*, at least within the time interval which is relevant for harvesting. The following relations exist:

(1) $\frac{w'(t)}{w(t)} > (<)$ M implies that $B'(t) > (<)$ 0, i.e. biomass weight increases (decreases).

(2) For $t = t_0$, where t_0 is defined by $w'(t_0)/w(t_0) = $ M so that $B'(t_0) = 0$, the biomass weight reaches its maximum. This occurs when the relative growth rate exactly equals the mortality rate.

Comparing the times for maximum individual weight and maximum yearclass weight reveals that

$$
t_0 < \tilde{t}.
$$

The biomass of fish – $B(t)$ – reaches its maximum before the individual fish – $w(t)$ – reaches its maximum. This is due to the fact that when $B(t)$ reaches its maximum, the individual growth rate is still positive $(w'(t_0)/w(t_0) = M > 0)$, while the growth rate is reduced to zero when the individual fish reaches its maximum weight $(w'(\tilde{t}) = 0)$.

3.2 BIOECONOMIC ANALYSIS

The biological model illustrates the changes in a yearclass of fish over time as a result of growth and natural mortality. Growth implies increased value for the fish farmer; mortality represents a loss. The purpose of this analysis is to find the harvesting time which maximizes the present value of the net revenues from the yearclass, within the given biological constraints.

The problem considered is an investment in fish and it is easily understood in the context of investment theory. Simply put, one wishes to maximize the present value of an investment by determining the optimal time of harvesting; hence only costs that influence the cash flow the investment generates are relevant. In the present case this amounts to variable costs such as feeding and harvesting costs. To simplify the analysis it will be assumed that the number of recruits is exogenously given; density dependence in the weight function is disregarded. In Appendix 1, determination of the optimal number of recruits is analysed.

The project analysed here concerns a *one-time* investment. What happens after the lifetime of the project is not considered. This problem is returned to in Section 3.4.

The analysis will be developed gradually, starting with a situation with no costs. Subsequently, different types of costs are introduced. Later, selective harvesting of fish is analysed.

Zero costs

The economic analysis begins with a hypothetical example of zero costs. This will give an intuitive understanding that can be a useful reference for more realistic models.

The *value* of the yearclass is found by multiplying price times quantity. Define

$$V(t) = p(w)B(t) = p(w)Re^{-Mt}w(t) \tag{5}$$

where $V(t)$ is gross biomass value and $p(w)$ is the price per kg of fish. The price is dependent on the size (weight) of the fish. Usually the price per kg is higher for large fish than for small fish $(p'(w) > 0)$.[2] There are, however, no seasonal variations in price. The number of fish released (R) and the growth curve are also considered exogenous variables. The development in $V(t)$ over time, which is derived by multiplying yearclass weight (Figures 3.3(a) and (b)) by the price, is illustrated in Figures 3.4(a) and (b). $t = t_{max}$ is the time of the maximum biomass value, i.e. $V'(t_{max}) = 0$.

If $p'(w) = 0$, i.e. the price of fish is not dependant on its weight, then

$$t_{max} = t_0.$$

In this case the time of maximum biomass value and the time of maximum biomass weight are the same. If $p'(w) > 0$, so that an increase in weight implies a higher price, then

$$t_{max} > t_0.$$

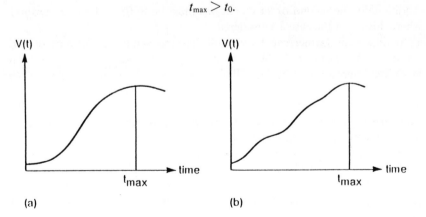

(a)   (b)

Fig. 3.4 (a) and (b) Biomass value $V(t) = p(w)B(t)$

One does, however, have $t_{max} < \tilde{t}$. Altogether, the following relations exist:

$$t_0 \leqslant t_{max} \leqslant \tilde{t}.$$

The biomass value reaches its maximum *value* at the same time or later than the time of maximum *weight* of the yearclass, as price depends on weight or not. However, the yearclass reaches its maximum value earlier than the time of maximum individual weight of the fish. At time $\tilde{t}$ the fish reaches its maximum individual weight, so that the value per fish is also greatest at that time. Due to natural mortality, the yearclass will reach its maximum value earlier.

For the case of zero costs, the fish farmer will harvest at the time which *maximizes the present value* of the biomass value[3] as considered at the time of releasing the fish:

$$\text{Max}_{\{0 \leqslant t \leqslant T\}} \quad \pi(t) = V(t)\, e^{-rt}.$$

Here, $\pi(t)$ is the present value of harvesting at time t, r is the interest rate, and T the life expectancy of the fish, or alternatively the time of sexual maturity. The harvesting time is the farmer's *control variable*. First order conditions for an optimum are found by taking the derivative with respect to time:

$$\pi'(t) = V'(t)e^{-rt} - rV(t)e^{-rt} = 0.$$

The optimal harvesting time, t^*, thus satisfies:

$$V'(t^*) = rV(t^*) \tag{6}$$

or

$$\frac{V'(t^*)}{V(t^*)} = r. \tag{6'}$$

Additionally the second order conditions must be fulfilled. Later an example where this is not the case is considered.

At time t the farmer can harvest all fish and acquire an income of $V(t)$. Assuming that the farmer's alternative is to deposit the money in a bank account at a given interest rate of r, this will give a return on investment equal to $rV(t)$. Thus $rV(t)$ becomes the *opportunity cost* of the fish farmer. On the other hand the farmer can refrain from harvesting at time t. The change in the biomass value over time is given by $V'(t)$, which will be the return on investment (in the form of fish, or capital in the sea). At the optimal time of harvesting, $t = t^*$, the return on investment of the capital deposited in the sea equals the alternative one on land. This can also be expressed as follows: the proportional increase in the biomass value equals the interest rate (Equation 6'). In the forestry literature this is known as the Fisher rule which denotes the optimal time of harvesting a single stand of trees.[4]

The following harvesting rules also illustrate the optimal rule:

(1) Do not harvest the fish if $V'(t) > rV(t)$. In this case the capital in the form of fish gives a better return on investment than one can obtain from a bank.
(2) Harvest the fish if $V'(t) \leqslant rV(t)$.

The optimal time of harvesting can also be illustrated graphically. For each point on the value curve, one can find the present value P by finding the P for which Pe^{rt} passes through this point on the value curve. Clearly, finding the $P*$ for which $P* e^{rt}$ is just tangent to the value curve also finds the highest present value of any point on the value curve. This is illustrated in Figure 3.5. The curve $P*e^{rt}$ is tangent to the value curve $V(t)$ for $t = t*$, which is the optimal time of harvesting. As the curves are tangents for $t = t*$,

$$\frac{d}{dt}\{P* e^{rt*}\} = \frac{d}{dt}\{V(T*)\}.$$

Differentiating obtains

Left hand side: $\dfrac{d}{dt}\{P* e^{rt*}\}$ $= rP* e^{rt*} = rV(t*)$

Right hand side: $\dfrac{d}{dt}\{V(t*)\}$ $= V'(t*)$

This of course gives the same condition for the optimal harvesting time as previously obtained in Equation (6).

In the 'traditional' Beverton–Holt model the harvesting time is determined unambiguously. If there are variations in the biomass growth and value curves (see Figures 3.3(b) and 3.4(b)), multiple solutions to the optimal harvesting time may be obtained. The case illustrated in Figure 3.6 may be realistic in fish farming. Here the value curve reaches a local maximum at time t_1, and a global maximum at time t_2. One can also imagine that there is more than one local maximum. For the two cases in Figure 3.6, both first and second order

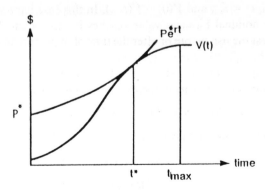

Fig. 3.5 Optimal harvesting time $t*$

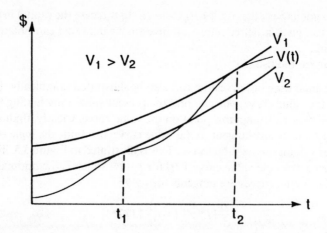

Fig. 3.6 Local maximum (t_1) and global maximum (t_2)

conditions for a maximum are satisfied at times t_1 and t_2. Hence the global maximum must be searched for. This question will not be pursued here, but note the implication that the Fisher rule is not foolproof.

Subsequently it shall be assumed that the optimal harvesting time, t^*, is unambiguously determined. For $r > 0$,

$$t^* < t_{max}.$$

This says that the optimal time of harvesting is less than the time that yields maximum biomass value (t_{max}). This is a result of discounting future incomes. If there is no discounting, i.e. $r = 0$, the optimal harvesting time is given by the condition:

$$V'(t^*) = 0.$$

In this situation, $t^* = t_{max}$ and $V(0) = V(t_{max})$. In this case harvesting will take place when the nominal biomass value reaches its maximum. Whether it is optimal to harvest the fish before or after the time of maximum biomass weight, t_0, depends on the values of the parameters.

It is possible to acquire a better understanding of the harvesting rule by evaluating the separate elements in the biomass value more closely:

$$V(t) = p(w)B(t) = p(w)Re^{-Mt}w(t).$$

Changes in $V(t)$ over time are given by:

$$\begin{aligned} V'(t) = \quad & p'(w)w'(t)Re^{-Mt}w(t) \\ & - Mp(w)Re^{-Mt}w(t) \\ & + w'(t)p(w)Re^{-Mt}. \end{aligned} \qquad (7)$$

The first component of this expression takes into account the increment in value due to increased weight $(w'(t))$ which causes the price to rise $(p'(w) > 0)$. The second component represents the economic loss resulting from the natural mortality of fish; the third component shows the increased value due to growth. The expression for $V(t)$ can be rewritten as:

$$V'(t) = \{\frac{p'(w)}{p(w)} w'(t) - M + \frac{w'(t)}{w(t)}\} V(t). \tag{7'}$$

The three components in brackets express the price appreciation due to growth, the natural mortality rate and the growth rate.

The rule for optimal harvesting says that the fish must be harvested when the marginal increase in the value of the 'natural' capital (i.e. fish in the sea) exactly equals the opportunity cost:

$$V'(t^*) = \left\{\frac{p'(w)}{p(w)} w'(t^*) - M + \frac{w'(t^*)}{w(t^*)}\right\} V(t^*) = rV(t^*), \tag{8}$$

which can be rewritten as

$$\frac{p'(w)}{p(w)} w'(t^*) + \frac{w'(t^*)}{w(t^*)} = r + M. \tag{8'}$$

The marginal revenue per fish with respect to time is $[p'(w)w(t) + p(w)w'(t)]$. Marginal user cost per fish is $[r+M]p(w)w(t)$. Equating marginal revenue and marginal user cost per fish and dividing by the value per fish $(p(w)w(t))$ yields Equation (8').

The fish must not be harvested when the capital (fish in the sea) gives a better return than the opportunity cost. This condition can be expressed as follows:

$$\frac{p'(w)}{p(w)} w'(t) + \frac{w'(t)}{w(t)} => r + M. \tag{9}$$

The left hand side expresses the *marginal revenue with respect to time*; relative growth rate $\left[\frac{w'(t)}{w(t)}\right]$ plus the price appreciation of the weight increase $\left\{\frac{p'(w)}{p(w)} w'(t)\right\}$. One assumes the marginal revenue to be decreasing over time.

The right hand side expresses the *marginal cost* by refraining from harvesting; the interest rate (r) plus the natural mortality rate of fish (M). The marginal cost is constant over time. The rule of optimal harvesting hence implies that one must refrain from harvesting the fish when the marginal revenue from waiting is greater than the marginal cost.

The optimal harvesting time, t^*, is shown in Figure 3.7. The marginal revenue curve $\{\frac{p'(w)}{p(w)} w'(t) + \frac{w'(t)}{w(t)}\}$ is declining over time, while the marginal cost $\{r + M\}$

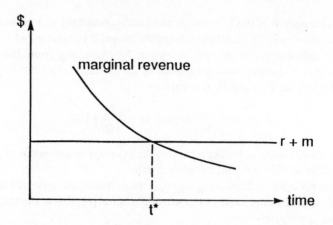

Fig. 3.7 The optimal harvesting time *t**

is constant and thus represented by a horizontal line. *t** is given by the intersection of the two curves.

An increase in the discount rate causes an upward shift in the marginal cost curve. As the marginal revenue curve with respect to time is unchanged while cost has increased, the optimal harvesting time is reduced. An increase in the natural mortality rate would have a similar effect on optimal harvesting.

Harvesting costs

Now the influence of costs on the harvesting time will be analysed. At first, harvesting costs will be considered. In its pure form this situation is more realistic for ocean ranching than for fish farming. One may envisage two types of harvesting costs: a cost that depends on the quantity harvested and a cost that depends on the number of fish harvested. These are not mutually exclusive and both will be considered here.

A harvesting cost per quantity unit (kg)

Assume that there is a fixed harvesting cost per kg fish of C_k, so that by harvesting all fish at time t the total harvesting costs will be $C_k B(t)$. The farmer faces the following maximization problem:

$$\begin{aligned}\text{Max} \atop \{0 < t \leqslant T\} \quad \pi(t) &= \{p(w)B(t) - C_k B(t)\}e^{-rt} \\ &= \{p(w) - C_k\}B(t)e^{-rt}\end{aligned}$$

Now $\{p(w) - C_k\}$ is the net price per kg the farmer receives when harvesting. The first order condition for a maximum is given by

$$\pi'(t) = p'(w)w'(t)B(t)e^{-rt} + \{p(w) - C_k\}B'(t)e^{-rt} - r\{p(w) - C_k\}B(t)e^{-rt}$$

Through simplification an implicit expression for the optimal harvesting time is derived:

$$\frac{p'(w)w'(t^*)}{p(w) - C_k} + \frac{w'(t^*)}{w(t^*)} = r + M. \tag{10}$$

Assuming $p'(w) > 0$, one notices that the price appreciation term is different compared to equation (9). Price appreciation is now on the basis of net price $\{p(w) - C_k\}$ rather than on the basis of gross price $p(w)$ as in Equation (9). Thus the marginal revenue curve with respect to time has shifted upwards and the optimal harvesting time has increased. For the case of $p'(w) = 0$, however, the price appreciation term vanishes. The optimal harvesting time is then independent of the per unit harvesting cost.

A harvesting cost per fish

Assume that there is a fixed harvesting cost per fish of C_s, so that by harvesting all fish at time t the total harvesting costs will be $C_s N(t)$.

As before, the farmer desires to maximize the present value of the income from his investment. The maximization problem is as follows:

$$\text{Max} \atop \{0 < t \leq T\} \quad \pi(t) = \{V(t) - C_s N(t)\}e^{-rt} = \{p(w)w(t) - C_s\}Re^{-(M+r)t}.$$

The first order condition for a maximum is given by

$$\pi'(t) = \{p'(w)w(t) + p(w)\}w'(t)Re^{-(M+r)t}$$
$$- (M+r)\{p(w)w(t) - C_s\}Re^{-(M+r)t} = 0.$$

Through simplification, the following expression for the optimal harvesting time is derived:

$$\frac{p'(w)}{p(w)}w'(t^*) + \frac{w'(t^*)}{w(t^*)} = [r + M]\left\{\frac{p(w)w(t^*) - C_s}{p(w)w(t)}\right\}. \tag{11}$$

As before one can deduce a harvesting rule; the fish should not be harvested as long as the capital (i.e. the fish) yields a better return than one can obtain alternatively:

$$\frac{p'(w)}{p(w)}w'(t) + \frac{w'(t)}{w(t)} > (r + M)\left\{\frac{p(w)w(t) - C_s}{p(w)w(t)}\right\}. \tag{12}$$

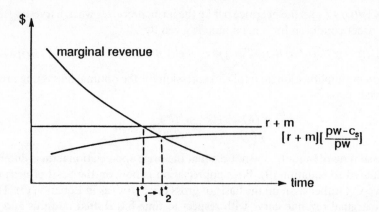

Fig. 3.8 Optimal harvesting time with and without harvesting costs

Compared to Equation (9), one notices that the marginal revenue by not harvesting the fish is unaltered. However, the *marginal cost* has changed as a result of the introduction of harvesting costs. $p(w) w(t)$ is the gross value per fish at time t, whereas $(p(w) w(t) - C_s)$ constitutes the net value. The new expression on the right hand side of the equation is the net value as a fraction of the gross value, which is less than one. Multiplied by $r+M$, this gives the marginal cost with respect to time, which in this way is reduced as compared to the original formulation of the problem.

This is understood intuitively by considering that the farmer has live fish as his capital. So far the value of the capital has been equal to the gross value of the fish. By introducing harvesting costs, the value of the capital *qua* fish is reduced. The marginal cost of refraining from harvesting the fish is consequently also reduced. It is hence optimal to wait a little longer before harvesting the fish than in the situation without harvesting costs.

The optimal harvesting time is illustrated in Figure 3.8. Without harvesting costs $t^* = t_1^*$, and with harvesting costs $t^* = t_2^*$ with $t_2^* > t_1^*$. As shown, harvesting costs imply that one must harvest later than otherwise. By postponing the harvesting a while, the discounted value of the harvesting costs will be reduced as a consequence of natural mortality. Discounting by itself also reduces the present value of harvesting costs. Further, an increase in the harvesting cost causes a downward shift in the marginal cost curve, thereby increasing the optimal harvesting time.

Feed costs

So far growth has been regarded as a function of time. This ignores the relationship between feeding and growth. In fish farming this assumption is unrealistic and it will be altered. The analysis will be developed gradually. The conversion ratio (f_t) is defined as follows:

$$f_t = \frac{F(t)}{w'(t)} \tag{13}$$

where $F(t)$ is the quantity of feed. The conversion ratio is thus the relation between the feed quantity and the growth of the fish. As a simplifying assumption, this factor is commonly set to be constant. The feed quantity per fish at time t is then:

$$F(t) = f_t w'(t). \tag{13'}$$

Note that the feed quantity varies over time according to the growth of the fish. More generally one could imagine that the feed quantity is dependant both on the growth rate $(w'(t))$ and the weight of the fish $(w(t))$. This is so because the fish need food not only to grow but also to retain body weight. However, this is disregarded in this analysis. Total feed quantity at a given time is

$$F(t) N(t) = F(t) Re^{-Mt} = f_t w'(t) Re^{-Mt}. \tag{14}$$

This expression is derived by inserting $N(t)$ from Equation (3'). The equation takes into account the loss of fish resulting from natural mortality.

The fish are fed continuously over time. Total feed quantity (SF_t) is found by *summing the feed quantity* from the time the fish are released until time t. This can be done mathematically by integrating Equation (14):

$$SF_t = \int_0^t F(u) Re^{-Mu} \, du. \tag{15}$$

From an economic point of view, feed costs are a concern. Let the price per unit (kg) of feed be C_f, which is constant over time. Multiplied by SF_t (Equation 15) this gives the total feed costs at time t. Discounted back to the time of releasing the fish, $t = 0$, one obtains:

$$\text{Present value of feed costs} = \int_0^t C_f F(u) Re^{-Mu} e^{-ru} \, du.$$

The farmer's maximation problem is as follows:

$$\text{Max } \pi(t) = V(t) e^{-rt} - \int_0^t C_f F(u) Re^{-(M+r)u} \, du.$$
$$\{0 \leqslant t \leqslant T\}$$

The first order condition for profit maximation is then:

$$\pi'(t) = V'(t) e^{-rt} - rV(t) e^{-rt} - C_f F(t) Re^{-(M+r)t} = 0.$$

By use of Equation (7') and through simplification, this can be rewritten as:

$$\frac{p'(w)}{p(w)} w'(t^*) + \frac{w'(t^*)}{w(t^*)} = r + M + \frac{C_f F(t^*)}{p(w) w(t^*)}. \tag{16}$$

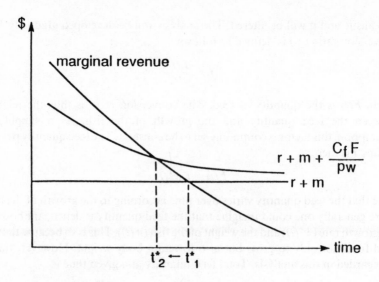

Fig. 3.9 Optimum harvesting time with and without feed costs

In comparison with the original first order condition for profit maximation (Equation 8'), the marginal feed costs must now be included in the opportunity cost. $C_f F(t)$ are the feed costs per fish at time t and $p(w)\,w(t)$ is the value of the fish. $C_f F(t)/p(w)\,w(t)$ is thus the relative feed cost, which combined with the interest rate and the rate of natural mortality, constitutes the cost of not harvesting the fish at time t. The determination of the optimal harvesting time is illustrated in Figure 3.9. In this situation the marginal revenue curve with respect to time is as before, but the marginal cost has increased as a result of the feed costs. This implies harvesting the fish earlier than before.

Feed and harvesting costs

When all costs are included, the following rule for the optimal harvesting time is derived:

$$\frac{p'(w)\,w'(t^*)}{p(w) - C_k} + \frac{w'(t^*)}{w(t^*)} = (r+M)\left\{\frac{p(w)\,w(t^*)-C_s}{p(w)\,w(t^*)}\right\} + \frac{C_f F(t^*)}{p(w)\,w(t^*)}. \quad (17)$$

While the feed costs imply that it is optimal to harvest the fish earlier than otherwise, the harvesting costs work in the opposite direction. The net result is an empirical question and varies from species to species.

Selective harvesting

One of the assumptions underlying this analysis is that all fish in the yearclass

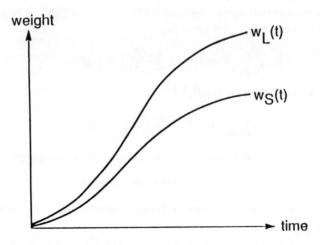

Fig. 3.10 Weight curves for large $(w_L(t))$ and small $(w_S(t))$ fish. $w_L(t) = aw_S(t)$, $a > 1$.

weigh the same. This has been done to simplify the analysis, without influencing the qualitative results. In reality the fish will vary in weight.[5] This is also the case in animal breeding and forestry. The problem is illustrated in Figure 3.10, with two weight curves for the same yearclass. In order to simplify the problem, imagine that there are only two types of fish – large or small. $w_L(t)$ shows the weight curve of large fish and $w_S(t)$ the weight curve of small fish. The question under consideration is whether to harvest all fish – small and large – at the same time or to harvest at different times. The latter can only take place if it is practically possible to separate large fish from small ones in the farm and the costs of doing so are not excessive.

As the question of selective harvesting is of considerable practical interest, a simple analysis of this problem is offered. All types of costs, including the ones related to separating small and large fish are disregarded. The relationship between the two sizes of fish is assumed to be:

$$w_L(t) = aw_S(t), \qquad a > 1.$$

It is assumed that the weight of the large fish is a constant factor of the weight of small fish during their whole lifetime; e.g. for $a = 1.5$, large fish always weigh 50 per cent more than small fish. Accordingly, the fry that are released also belong to two weight classes.

The general formula for optimal harvesting is given by Equation (8'):

$$\frac{p'(w)}{p(w)} w'(t^*) + \frac{w'(t^*)}{w(t^*)} = r + M.$$

All parameters are assumed to be equal for both types of fish, except for the weight curves. As this problem really concerns two different types of fish in the

same farm, the optimal harvesting time for both types of fish can be found:
Optimal harvesting time for *large fish*:

$$\frac{p'(w)}{p(w)} aw_S'(t^*) + \frac{w_S'(t^*)}{w_S(t^*)} = r + M.$$

Optimal harvesting time for *small fish*:

$$\frac{p'(w)}{p(w)} w_S'(t^*) + \frac{w_S'(t^*)}{w_S(t^*)} = r + M.$$

The following relationship between the weight curves has been used:

$$w_L(t) = aw_S(t) \Rightarrow w_L'(t) = aw_S'(t).$$

This implies the following relationship between growth rates – $w'(t)/w(t)$ – of the two types of fish:

$$\frac{w'_L(t)}{w_L(t)} = \frac{aw_S'(t)}{aw_S(t)} = \frac{w_S'(t)}{w_S(t)}.$$

This expression shows that relative growth is equal for both large and small fish. This is a result of the relationship between the two weight curves.

For large and small fish the marginal costs with respect to time – $r+M$ – are identical. It is observed that relative growth is also equal. The question of selective harvesting hence depends on whether the component

$$\frac{p'(w)}{p(w)} w'(t),$$

i.e. the price appreciation due to growth, differs or not. So far we have assumed that $p'(w) > 0$. One can distinguish three cases:

(1) $p'(w) = 0$, i.e. the price per kg of fish is constant irrespective of size. If so, there is no price appreciation as a result of growth and this component equals zero. The marginal revenue with respect to time is then determined by $w'(t)/w(t)$, i.e. relative growth. As this is equal for both types of fish, it will be optimal to harvest all fish at the same time.
(2) $p'(w) > 0$, i.e. the fish price increases with weight. This is the case with salmon. Due to a greater weight increase in absolute terms for large fish than for small, a relatively greater price appreciation occurs for large fish. The marginal revenue with respect to time is greater for large fish than for small, while the marginal cost is the same. The result of this is that one should harvest small fish earlier than large fish.
(3) $p'(w) < 0$, i.e. large fish – within a relevant time interval regarding harvesting – are worth less per kg than small fish. In this situation it will be profitable to harvest large fish before small fish.[6] This refers to a

situation where small fish are preferred for a specific usage, so that small fish have a relatively higher value.

Here only two types of fish have been considered: small or large. It is obvious that the results can be generalized to situations with several sizes of fish. The necessary assumption is again that it is actually possible to separate the fish. In the problem formulation, the relative growth rates for all types of fish are equal. Absolute weight increases vary, however, which results in differing price appreciations of the fish. When this is the case, selective harvesting will be of interest.

Another possible situation is that of *differing growth rates* – $w'(t)/w(t)$ – for different fish of the same age, at least within relevant time intervals. This formulation is more realistic than the former one. Assuming the growth rate of large fish to be greater than that of small fish, the problem can be analysed with reference to Equation (8'). The left hand side of this equation can be rewritten as

$$\left\{ \frac{p'(w)\,w(t)}{p(w)} + 1 \right\} \frac{w'(t)}{w(t)}.$$

The optimal harvesting time now depends on how the elasticity $p'(w)\,w(t)/p(w)$ changes with $w(t)$.

If $p'(w) = 0$, the marginal revenue with respect to time equals the relative growth rate. This decreases over time. However, as the growth rate of large fish is greater than that of small fish, one must harvest small fish before large. On the other hand, if price is not constant, the change in the elasticity depends on both the first and second derivatives of $p(w)$. To investigate this matter, one would have to make assumptions about the curvature properties of the price function. This issue will not be pursued here.

In the first case, price appreciation alone could cause selective harvesting to be optimal. In the second and more general case, differences in relative growth rates can give rise to selective harvesting. These conclusions are not altered if feed and harvesting costs are put into the model. The expressions for the optimal time of harvesting will then be modified as above.

The conclusion is that selective harvesting should be considered in situations where the growth of the fish of the same yearclass differs. The practical consequence of this conclusion depends on whether it is possible to separate fish of various sizes and the costs of doing so.

3.3 THE ROTATION PROBLEM

The optimal time for harvesting a yearclass of fish has been analysed. The analysis did not consider that when the fish are actually harvested, space is made

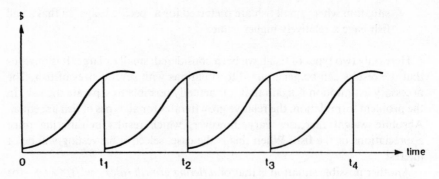

Fig. 3.11 The rotation problem

available for new fish. This is an important aspect as the space (volume) in fish farms is limited. As the marginal value decreases over time, harvesting can make room for new fish which may grow faster and thereby yield a greater increase in value. Therefore, it is not sufficient to merely consider a single harvesting time. A sequence of such times must be determined. This is illustrated in Figure 3.11.

This problem may be clarified with reference to investment theory. So far, a one-time investment in fish has been considered without taking into account what happens to the project when its lifetime has expired. When the fish are harvested space is made available for releasing new fish. These then represent a new investment. Assuming the production capacity to be constant into the future, the problem represents an *indefinite* series of indentical investments and not a one-time investment. This aspect will now be included in the analysis.

All parameters are assumed to be constant over time. The problem then is reduced to finding the optimal *rotation time* for a yearclass of fish, i.e. the time of harvesting a yearclass and releasing the next.

Consider a sequence of times

$$t_2 < t_2 < t_3 < \ldots$$

with the property that at time t_k the fish will be harvested and new fish will be released. $t = 0$ is the time of releasing the first yearclass. This is the same as the first investment in fish. The present value (PV) at time $t = 0$ of all future incomes is given by

$$PV = V(t_1)e^{-rt_1} + V(t_2)e^{-rt_2} + V(t_3)e^{-rt_3} + \ldots \tag{18}$$

where r is the interest rate. Equation (18) is solved for the optimal rotation time in Appendix 1. Rotation is found to imply a reduction in the optimal harvesting time. This is because the farmer can replace slower growing old fish with faster growing young fish.

In this analysis it is assumed that when one yearclass is harvested, the next one is released immediately. This implies that 'recruits' are available throughout the year. In reality this is not so for all species. Salmon spawn only at certain times. This determines when new smolts become available and contradicts the assumption on which the formulation of the rotation problem is based. However, with other species, such as turbot, it is possible to control the time of spawning so that fry are available for the whole year. The rotation problem is therefore relevant for farming these species.

The situation for salmon may also change over time as more effort is put into controlling the spawning time. There have already been fairly successful experiments with releasing smolts in autumn. Moreover, in land-based farms one can better control temperature and other parameters than in sea-based farms. Then growth is not so heavily dependent on natural temperature conditions. Under these conditions the rotation problem as formulated here will be of relevance for salmon farming. This may imply a competitive advantage for land-based plants.

The rotation problem may also be made relevant for salmon by reformulating the model slightly. Smolts are now in principle available only once a year. By formulating the rotation problem in discrete time, this model can determine in what year to harvest the fish. (For instance, one or two years after releasing them.) This permits the use of continuous time analysis for determining the time in that year for harvesting.

Today fish are mainly harvested in the second year after being released. There is, however, reason to believe that some market segments desire salmon of 2–3 kg, approximately one-year old salmon. Harvesting in the first year may therefore become a realistic alternative to harvesting in the second year. An economic analysis of this must be based on a discrete form of the model which is presented here.

3.4 OPTIMAL HARVESTING: EXAMPLES

This section provides examples of optimal harvesting for two species, salmon and turbot. Previously the effect of different costs on harvesting time has been analysed. Now through numerical examples some quantitative results are obtained.

Examples of weight curves are given by:

(1) $w(t) = w_\infty \left(a - be^{-ct}\right)^3$

(2) $w(t) = e^{a-b/t}$

(3) $w(t) = w_0 + a_1 t + a_2 t^2 + a_3 t^3$

(4) $w(t) = \dfrac{w_\infty}{1 + a\,e^{-bt}}$

The first function is von Bertalanffy's weight function, where w_∞ is the maximum individual weight fish will reach asymptotically over time. This function has been used for cod fisheries and could constitute a basis for an analysis of ocean ranching of cod. In the second case, the fish will grow towards an asymptotical value (e^a), while the third one is given by a third degree polynomial. The last case is given by a logistic function where the fish also grow towards an asymptote (w_∞).

Weight curves vary from species to species and from location to location, depending on several factors. The results which are represented here must therefore only be considered as examples. In spite of this, there is reason to believe that certain qualitative characteristics of the results may be generalized.

The data for estimating the weight curves are given in Appendix 2.

Salmon

The following weight curve has been estimated, based on growth observations for salmon:[7]

$$w(t) = 2.64t^2 - 0.74t^3.$$

$w(t)$ is the weight of the fish at time t, measured in years from the time they were released. For this set of data a third degree polynomial, with the constant and the first degree component set equal to zero, gave the best result.

The estimated growth curve is illustrated in Figure 3.12, where the observed

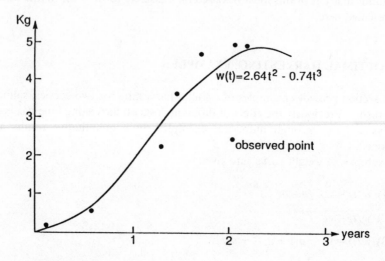

Fig. 3.12 Weight curve for salmon

data points are also shown. The following assumptions have been made about natural mortality and price:

$$M = 0.1$$
$$p(w) = 28.90 + 2.87 \, w(t)$$

This implies that approximately 10 per cent of the fish will die naturally in one year.[8] The price is in Norwegian Kroner (NOK) per kg.[9] With these parameter values one obtains

$$\tilde{t} = 2.36 \text{ years with}$$
$$w(\tilde{t}) = 4.9 \text{ kg,}$$
$$t_0 = 2.28 \text{ years, and}$$
$$t_{max} = 2.30 \text{ years.}$$

According to the weight curve, the fish reaches its maximum weight – 4.9 kg – at 2.36 years $(\tilde{t})$ after being released. The yearclass, however, reaches its maximum weight after 2.28 years $(\tilde{t}_0)$ As the fish price increases with the weight of fish, the maximum biomass value is reached later. Here, after 2.30 years.

The following assumptions have been made about costs:[10]

Feed price per kg:	C_f = NOK 6.00
Conversion ratio:	f_t = 1.70
Harvesting cost per fish:	C_s = NOK 15.00
Insurance premium:	k = 0.04

Table 3.1 gives the optimal time of harvesting salmon for different cost alternatives and interest rates.[11] In the zero cost alternative, fish are to be harvested at the time of maximum biomass value $V(t)$, i.e. $t^* = t_{max}$ for $r = 0$.

There are two conclusions to draw from the results. First, the optimal harvesting time is relatively insensitive to changes in the interest rate. An increase in r from 0.1 to 0.2 reduces t^* with about half a month. Second, the harvesting time is only to a small extent influenced by variable costs. Note that the

Table 3.1 Optimal Harvesting Time (t*) in Years for Salmon

r Interest rate	t^* Zero costs	t^* Harvesting costs	t^* Feed and insurance costs	t^* Feed, insurance and harvesting costs
0.00	2.305	2.310	2.260	2.265
0..05	2.265	2.275	2.215	2.225
0.10	2.230	2.240	2.170	2.180
0.15	2.190	2.205	2.125	2.140
0.20	2.150	2.165	2.080	2.095

harvesting costs imply an increase in t^*, but feed and insurance costs imply a reduction in t^*. This corresponds to the analytical results of Section 3.3. In the case considered here, the feed and insurance costs dominate. The net result is that the costs imply a decrease in t^* compared to a situation with no costs.

A change in the natural mortality rate (M) will have an identical effect on the optimal harvesting time as an equivalent change in the interest rate. This is clearly indicated in Equation (8′). (The same is true for the insurance rate; see Appendix 1).

Turbot[12]

The same type of weight curve was estimated for turbot as for salmon. This gave the following result:[13]

$$w(t) = 0.78t^2 - 0.18t^3.$$

As before, $w(t)$ is the weight of the fish, with time measured in years from the time of release.

The weight curve is shown in Figure 3.13 where the observed data points are also marked. The following assumptions are made:

$$M = 0.1$$
$$p = \text{NOK } 50.00 \text{ per kg.}$$

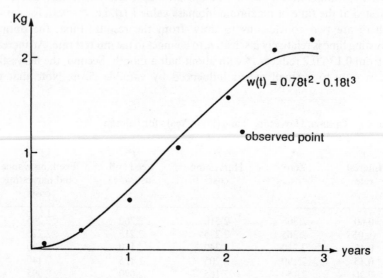

Fig. 3.13 Weight curve for turbot

This gives

$$\tilde{t} \quad = 2.84 \text{ years with}$$
$$w(\tilde{t}) = 2.1 \text{ kg}$$
$$t_0 \quad = t_{max} = 2.7 \text{ years.}$$

The turbot reaches its maximum weight of 2.1 kg 2.84 years after it is released $(\tilde{t})$.[14] With $M = 0.1$, the yearclass reaches its maximum weight after 2.7 years (t_0). As price is independent of weight, this also equals the time of maximum biomass value. The following assumptions have been made concerning costs:

Feed price per kg: $C_f =$ NOK 4.00
Conversion ratio: f_t = 2.80.

Harvesting and insurance costs are disregarded.

The optimal harvesting time is given in Table 3.2. An increase in r from 0.1 to 0.2 causes a reduction in t^* of about two months. The harvesting time for turbot is more sensitive to changes in the interest rate than for salmon. The time rate of change in value for turbot decreases more rapidly than for salmon.

The optimal rotation time is also estimated (Table 3.3). Optimal rotation of turbot implies that one must harvest the fish 0.7–0.8 years earlier than in a situation where the harvested fish are not replaced with new fish. As the weight

Table 3.2 Optimal Harvesting Time (t*) in Years for Turbot

Interest rate	t^* Zero costs	t^* Feed costs
0.00	2.70	2..66
0.05	2.63	2.56
0.10	2.55	2.46
0.15	2.47	2.36
0.20	2.39	2.26

Table 3.3 Optimal Rotation Time in Years for Turbot

Interest rate	Rotation time No costs
0.00	1.905
0.05	1.850
0.10	1.795
0.15	1.740
0.20	1.685

curve of turbot is based on data from a land-based installation, a further analysis of this question would be of interest.

3.5 POSTSCRIPT

This chapter has analysed optimal harvesting of farmed fish. The basic assumption has been that the physical limitations have been given, e.g. by investments in production capacity in fish farms or by cordoning off a fiord for sea ranching. The analysis started by introducing a simple biological model. This was used as a basis for later extensions. Additionally, examples were given for practical use in the analysis of two species.

The theoretical analysis has a stronger resemblance to optimal exploitation of forestry than traditional fisheries. This is because fish farming takes place under relatively controlled conditions, where the fish farmer owns the fish. It is easily seen that this can be extended to a question of optimal harvesting in forestry.

The chapter is for the most part an application of theory to a new problem area, and two new results seem to have emerged from the analysis. One is that harvesting costs – exemplified by a unit cost per fish – imply that one should harvest the fish later than in a situation without such costs. The second is that differences in growth can cause selective harvesting to be optimal. Both these results are applicable to forestry.

The point of reference for the analysis was a fish farmer who considered the fish price as given. Introducing a price function, with the fish price varying over time, will not lead to any qualitative changes in the analysis. One associated problem is that if many farmers follow the same harvesting rule, there might at a given time be such a large supply of fish that the price drops. Then the actual price obtained will differ from the price the farmer used to calculate the optimal harvesting plan. To some extent this will be offset by differences in growth conditions from farm to farm, which will cause the optimal harvesting time to vary. In addition, selective harvesting may result in a certain smoothing out of deliveries over time. However, this is probably not sufficient to solve the problem raised. An extension of the analysis to include this problem would be of interest.

The analyses that have been presented here can be employed in practical fish farming through discretizing the model on a monthly basis. This will be done in the next chapter. It will also be easy to introduce types of costs other than the ones considered here.

NOTES

1 u is used as an integration symbol rather than t, to avoid misunderstandings.
2 As weight $w(t)$ is a function of the *age* of the fish, t, price is also a function of age.

3 The interest rate is here expressed as *continuous time* interest rate. See Appendix 1 for the relationship between continuous and discrete time interest rates.

4 This harvesting rule is named after the economist Irving Fisher.

5 Density may also influence growth. This will be disregarded in the following example.

6 The price per kg may increase up to some limit but then decrease for fish that exceed this limit. In this case one would presumably harvest fish before they reach the size at which price starts declining. This will also depend on the growth rate.

7 The weight curve is estimated by using the least squares method. t statistics for the two parameter estimates are 15.04 and –9.27 respectively, with an adjusted $r^2 = 0.95$ and Durbin–Watson statistic of 1.87. When a constant term was included in the regression, the parameter estimate was negative and insignificant. Excluding the constant had a negligible effect on the other coefficients.

8 This parameter value gives a total natural mortality of more than 20 per cent over the life time of the fish. This is not unreasonable.

9 This price function is a linear approximation to ex-farm salmon prices.

10 The cost figures are based on experience in the industry; see also Chapter 5.

11 See Appendix 1 for influence of insurance costs on the optimal harvesting time.

12 The parameter values for turbot – except for the weight curves – have been established in association with Mr J. Stoss, biologist at Øye Havbruk, a turbot producer.

13 t statistics for the two parameter estimates are 37.53 and –26.68 respectively, with $r^2 = 0.99$.

14 According to the available data, the weight of turbot stabilizes at approximately 2 kg. The turbot can, however, grow substantially larger, but the data do not give the basis for analysing this further.

4 A harvesting model for a fish farm

This chapter analyses production planning in fish farms through an analysis of optimal harvesting for salmon. The analysis makes use of a discrete time model which is updated once a month with respect to important variables such as the number of fish, growth, mortality and feeding. It is based on the continuous time model of Chapter 3. Apart from a different definition of time, the two models make use of the same basic method. It should be noted, however, that the monthly model is easier to use for fish farms.

The principles of cash flow analysis are analysed in Section 4.1, as these constitute the basis for the analysis. An *example* of optimal harvesting in fish farms is considered in Section 4.2.

The primary intent of this chapter is to give a practical illustration of the *method* employed in optimal production planning. Due to the many calculations involved, sensitivity analyses for changes in parameter values are not done. For actual planning in fish farms, personal computer models should be used.

4.1 CASH FLOW ANALYSIS

The starting point is a fully developed fish farm. Here the optimal operation of fish farms is analysed. Only one yearclass of fish is considered. In this way all the major results are derived at a level that is readily understood. The analogy to a case with several yearclasses should also be clear.

The reference point for the analysis is the time of smolt release. The farmer's investment in fish is considered, the objective being to find the harvest time which maximizes the present value of the investment. The following assumptions constitute the basis of the analysis:

(1) The plant is fully developed. All fixed costs related to plant operation are disregarded.
(2) Labour is primarily a fixed cost.
(3) Smolts have been purchased and released. Therefore only the harvesting time with respect to the number of smolts already released is maximized.

(4) What happens after the yearclass has been harvested (rotation) is not considered, i.e. only a one-time investment is considered.
(5) Credit is not a limiting constraint with respect to operations and taxes are disregarded.

From these assumptions it can be seen that the objective is to maximize the present value of cash flows from the investment; only variables that are directly affected by the investment are of relevance.

In addition to biological variables, the following economic variables are considered:

- Price
- Interest rate
- Feed costs
- Harvesting costs

The harvesting costs include the hiring of additional labour during the harvesting season. The analysis can easily be extended to include other variable costs.

The principal differences between the various types of costs are stressed. Feed costs are incurred every month, but harvesting costs are only incurred in the month(s) with actual harvesting. The model will take this into account.

The analysis is in terms of the cash flow the investment generates with respect to both revenues and expenditures. It is based on monthly periods. Let

V_t = revenue when harvesting in month t,
C_t = expenditures month t,
r = the interest (discount) rate per month.

With reference to the time of releasing the fish, $t = 0$, it is assumed that the fish farmer will harvest before the fish reach sexual maturity. Assuming this occurs after 29 months, the following alternatives with respect to harvesting exist:

	$t = 0$	$t = 1$	$t = 2$	...	$t = 29$
Sales revenue	V_0	V_1	V_2		V_{29}
Expenditures	C_0	C_1	C_2		C_{29}
Net cash flow	KS_0	KS_1	KS_2		KS_{29}

The problem facing the fish farmer is harvesting the fish in the month that maximizes the present value of the investment. Assuming all fish will be harvested in the same month, there are 30 possible investment alternatives; i.e. harvesting in the months zero to 29. In fact only harvesting in a shorter period will be realistic, during the last 12 months for example. More on this later.[1]

KS_t $(t = 0,1, \ldots 29)$ shows the *net cash flow* in nominal dollars if harvesting takes place in this month; i.e. sales revenue minus expenditures in the month of harvesting. In addition, expenditures in previous months must be taken into account.[2] In order to find the present value of the investment, the cash flows must be discounted back to the time of releasing the fish. There are the following alternatives for the investment in fish:

Harvest t = 0

$$\text{Present value} \quad PV_0 = KS_0 = V_0 - C_0$$

Harvest t = 1

$$\text{Present value} \quad PV_1 = \frac{KS_1}{1+r} - C_0 = \frac{V_1}{1+r} - \left\{ C_0 + \frac{C_1}{1+r} \right\}$$

Harvest t = 2

$$\text{Present value} \quad PV_2 = \frac{KS_2}{(1+r)^2} - \left\{ C_0 + \frac{C_1}{1+r} \right\}$$

$$= \frac{V_2}{(1+r)^2} - \left\{ C_0 + \frac{C_1}{(1+r)} + \frac{C_2}{(1+r)^2} \right\}$$

Harvest t = n; n = 0 ... 29

$$\text{Present value} \quad PV_n = \frac{KS_n}{(1+r)^n} - \left\{ C_0 + \frac{C_1}{1+r} + \frac{C_2}{(1+r)^2} + \ldots + \frac{C_{n-1}}{(1+r)^{n-1}} \right\}$$

$$= \frac{KS_n}{(1+r)^n} \quad - \quad \sum_{t=0}^{n-1} \frac{C_t}{(1+r)^t}$$

$$= \underbrace{\frac{V_n}{(1+r)^n}}_{\substack{\text{Present value} \\ \text{sales revenue}}} \quad - \quad \underbrace{\sum_{t=0}^{n} \frac{C_t}{(1+r)^t}}_{\substack{\text{Present value} \\ \text{expenditures}}}$$

As can be seen, it is necessary to discount the cash-flows – for revenues and expenditures – separately. This is so because there is sales revenue only in the month of harvesting. If only the net cash flow is discounted, feed and other costs incurred in all the other months will not be included.

By calculating the present value of harvesting in all possible months, the month which maximizes the present value can be found. This is analogous to the maximization in the continuous time analysis in the previous chapter.

4.2 OPTIMAL HARVESTING

This section provides an *example* of optimal harvesting of fish. The example is for *chinook* salmon. This will be done by considering a yearclass of fish and the cash flow it generates. It is assumed that revenues are obtained in the same month as harvesting takes place and that expenditures are paid in the month incurred. Hence problems relating to periodizing revenues and expenditures are disregarded.

All prices are expressed in nominal Canadian dollars (1989). Data are taken from British Columbia, Canada.

Biomass weight

This analysis is based on the following assumptions concerning plant operations and development of the yearclasses:

- In the month of April, year zero, 95 000 chinook salmon smolts are released. This means the month of April represents $t = 0$.
- Natural mortality: 1.5 per cent per month.

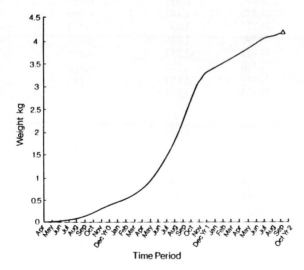

Fig. 4.1 Weight curve for chinook salmon

The *weight curve*, on which estimates are based, is illustrated in Figure 4.1. In this example the fish reach a maximum weight of about 4.2 kg in October of the second year. With the other assumptions concerning time of release and natural death, the development of the yearclass over time can be described. This is done

Table 4.1 Development of Biomass

Month	Number of fish N_t	Weight per fish (kg) w_t	Weight-increase (kg)[a] $w_{t+1} - w_t$	Biomass-weight (tonnes) $B_t = N_t w_t$
Year 0				
April	95 000	0.007	0.007	0.67
May	93 575	0.014	0.013	1.31
June	92 171	0.027	0.021	2.49
July	90 788	0.048	0.037	4.36
August	89 426	0.085	0.060	7.60
September	88 084	0.145	0.087	12.77
October	86 763	0.232	0.072	20.13
November	85 461	0.304	0.073	25.98
December	84 179	0.377	0.070	31.74
Year 1				
January	82 916	0.447	0.083	37.06
February	81 672	0.530	0.090	43.29
March	80 447	0.620	0.134	49.88
April	79 240	0.754	0.181	59.75
May	78 051	0.935	0.232	72.98
June	76 880	1.167	0.280	89.72
July	75 727	1.447	0.359	109.58
August	74 591	1.806	0.447	134.71
September	73 472	2.253	0.474	165.53
October	72 370	2.727	0.422	197.35
November	71 284	3.149	0.189	224.47
December	70 215	3.338	0.104	234.38
Year 2				
January	69 162	3.442	0.106	238.06
February	68 125	3.548	0.100	241.71
March	67 103	3.648	0.113	244.79
April	66 096	3.761	0.113	248.59
May	65 105	3.874	0.120	252.22
June	64 128	3.994	0.119	256.13
July	63 166	4.113	0.035	259.80
August	62 219	4.148	0.013	258.08
September	61 286	4.161	0.044	255.01
October	60 367	4.205	0.000	253.84

[a] This column shows the change in weight per fish from one period to the next.

in Table 4.1 for the *number of fish (N_t)* and *biomass weight (B_t)*. As noted above, it is assumed that the fish are released in April. The number of fish is derived by taking natural mortality into account. The third and fourth columns show the weight per fish *(w_t)* and the weight increase per fish *(w_{t+1} − w_t)*, respectively, cf. Figure 4.1.

The last column shows the biomass weight, which reaches a maximum in *July* in the second year after smolts are released. The biomass weight is then about 260 t, which constitutes the maximum harvesting quantity the yearclass can yield. After this time, natural mortality is greater than the weight increase so that the net change in biomass weight is negative. It should be noted that the time of maximum biomass weight (July year two) occurs before the time of maximum weight per fish (October year two). See Section 3.1 for a more elaborate explanation of the underlying biological model.

Sales revenue

It is assumed that all fish are harvested and paid for in the same month. Thus, the *revenue* will equal *biomass value (V_t)* in the month of harvesting:

$$V_t = p(w)B_t, \qquad t = 0, 1, \ldots, 29,$$

where B_t is biomass weight in month t and $p(w)$ the price. It is assumed that the price per kg *(p)* depends on the weight per fish, with large fish commanding a higher price per kg than small fish.

The development of biomass value, V_t, for the period July year one to October year two is shown in Table 4.2. As the price per kg salmon depends on the weight of the fish, the fish are divided into different size groups with price generally increasing with weight. The correct way of finding the biomass value at a point in time is to find the size distribution of the fish and calculate the value by taking into account the fact that prices vary with size. This has been done for the case considered here. The price column in Table 4.2 represents *average price* for each month in question.[3]

An alternative to this procedure is to use the average weight of the fish as shown in Table 4.1, and then to use the kg price in the respective size categories. The disadvantage of this approach is that when the average weight from one month to the next moves from one size category to another – for example from 2.70 to 2.75 kg, i.e. moving average size from the 4–6 lb range to the 6–9 lb range – this might result in an overstatement of the increase in value. When the mean weight equals 2.75 kg, 50 per cent of the fish are under and 50 per cent are over 2.75 kg. This means that nearly half the fish are in the less than 6 lb category, while the rest belong to the 6 lb and over category. In this example the kg price for 6–9 lb fish cannot be used to assess biomass value.

With the assumptions made here, average price increases moderately from

Table 4.2   Development of Biomass Value

Month	Price per kg (Cnd. $)	Biomass Value V_t (nominal) V_t	Biomass-value (discounted) $V_t/(1+r)^t$
Year 1			
July	5.06	554 459	477 583
August	5.60	754 384	643 354
September	6.36	1 052 786	888 949
October	7.11	1 403 180	1 173 083
November	7.60	1 705 997	1 412 122
December	7.74	1 814 083	1 486 722
Year 2			
January	7.87	1 873 498	1 520 212
February	7.94	1 919 158	1 541 844
March	8.02	1 963 230	1 561 635
April	8.09	2 011 069	1 583 850
May	8.19	2 065 655	1 610 733
June	8.26	2 115 611	1 633 353
July	8.34	2 166 747	1 656 270
August	8.37	2 160 167	1 634 891
September	8.47	2 159 944	1 618 537
October	8.50	2 157 667	1 600 823

month to month as weight increases. When the average price goes from one weight group to another, as in the example above, a big price jump is avoided and only a small increase occurs.

If, for example, September of year one is considered, the fish have reached an average weight of 2.25 kg. The average price per kg is $6.36. Total biomass weight equals 165.53 t, giving a potential sales revenue or biomass value of $1.053 million. Discounted back to time zero at one per cent per month, this equals $0.889 million. From Table 4.2 it is noted that the nominal biomass value reaches a maximum in July of year two, i.e. in the same month as maximum biomass weight.[4]

The discounted biomass value is illustrated in Figure 4.2. It increases substantially in the autumn of year one, then flattens out in the spring of year two. The discounted biomass value reaches its maximum in July of year two, the same time maximum nominal biomass value is reached.

Costs

It is important to distinguish between harvesting and feed costs. This is because

DISCOUNTED BIOMASS VALUE

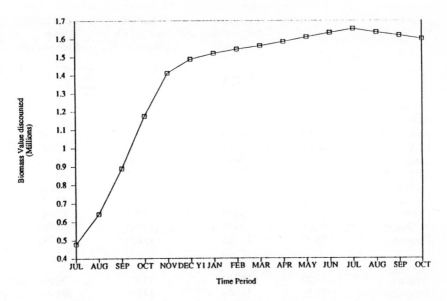

Fig. 4.2 Discounted biomass value from July of Year One to October of Year Two

harvesting costs are incurred only in the actual month of harvesting while the fish are fed throughout their lifespan.

As feed costs are incurred every month over the lifetime of the fish, they must be summed and discounted back to the time when the fish are released. The following assumptions are made:

- Conversion ratio: 1.7 kg feed per kg of weight increase
- Feed price: $1.30 per kg

The following relations exist:

(1) Feed consumption per month = weight increase x conversion ratio.

Expressed in symbols, this equals:

$$F_t N_t = (w_{t+1} - w_t) f N_t,$$

where F_t is feed consumption per fish per month, $(w_{t+1} - w_t)$ is the weight increase per month and f the conversion ratio, which is assumed to be constant. N_t is as before the number of fish. It is seen that feeding depends on the weight increase, the conversion ratio and the number of fish.

(2) Feed costs per month (nominal) = feed consumption x feed price.

Table 4.3 Feed Costs

Month	Feed quantity (tonnes) $F_t N_t$	Feed cost per month (nominal) $C_f F_t N_t$	Feed cost per month (discounted) $\dfrac{C_t F_t N_t}{(1+r)^t}$	Total feed costs (discounted) $\displaystyle\sum_{(=)}^{n} \dfrac{C_t F_t N_t}{(1+r)^t}$
Year 0				
April	1.13	1 469	1 469	1 469
May	2.07	2 691	2 664	4 133
June	3.13	4 069	3 989	8 122
July	5.71	7 423	7 205	15 327
August	9.12	11 856	11 393	26 720
September	13.03	16 939	16 117	42 837
October	10.62	13 806	13 006	55 843
November	10.61	13 793	12 865	68 708
December	10.02	13 026	12 029	80 737
Year 1				
January	11.70	15 210	13 907	94 644
February	12.36	16 068	14 546	109 191
March	18.33	23 829	21 358	130 549
April	24.38	31 694	28 127	158 676
May	30.78	40 014	35 159	193 835
June	36.59	47 567	40 972	234 806
July	46.22	60 086	51 243	286 049
August	56.81	73 853	62 983	349 032
September	59.08	76 804	64 852	413 884
October	52.04	67 652	56 558	470 442
November	22.90	29 770	24 642	495 084
December	12.41	16 133	13 222	508 306
Year 2				
January	12.58	16 354	13 270	521 576
February	11.58	15 054	12 094	533 670
March	12.89	16 757	13 329	546 999
April	12.70	16 510	13 003	560 002
May	13.28	17 264	13 462	573 464
June	13.08	17.004	13 128	586 592
July	3.76	4 888	3 736	590 328
August	13.75	17 875	13 528	603 857
September	4.58	5 954	4 462	608 318
October	0.00	0	0	608 318

This can also be expressed with symbols:

Feed costs per month: $C_f F_t N_t$,

where C_f is the feed price per kg. By summing the feed costs over the lifetime of the fish and discounting back to the time of release, the following expression is derived:

Present value of feed costs: $\qquad \displaystyle\sum_{t=0}^{n} \frac{C_f F_t N_t}{(1+r)^t}$

The calculations are illustrated in Table 4.3.

In order to find feed quantity per month, fish growth must be evaluated. This is given in the fourth column of Table 4.1. As regards the number of fish, mortality must be taken into account as some fish will die every month. When the fish actually die is of importance. The assumption here is that this occurs at the end of the month. Fish will therefore be fed the whole month in which they die. In reality this implies a slight overestimation of feeding costs.

In April of year zero, 95 000 smolts are released with an individual weight of 7 g. In May the weight has increased to 14 g, so that the weight increase is 7 g. With a conversion ratio of 1.7, 11.9 g of feed per fish is required. All 95 000 fish are actually fed in April, so the total feed quantity equals 1.13 t. In Table 4.3 this is shown under the month of April. The feed costs of $1,469 (nominal) are paid in April. As April is defined as time zero, there is no discounting for this month.

If, however, February of year two is taken as the point of reference, the weight increase per fish during this month is 0.1 kg. A feed factor of 1.7 implies a feed quantity of 0.17 kg per fish. On February 1, there were 68 125 fish in the pens. All these are fed this month, including the ones that die. Total feed quantity then is 11.58 t and, with a feed price of $1.30 per kg, the feed costs are $15 054.00. Discounting back to the time of release, these costs amount to $12 094.00. The last column of Table 4.3 shows the sum of the feed costs discounted back to the time of release. In February year two the total feed costs (discounted) are $533 670.00.

With respect to the other costs, the following assumption is made:

Harvesting costs: $3.10 per fish.

Harvesting costs are assumed to depend on the number of fish harvested and not the biomass weight of fish (see Section 3.2 for a theoretical analysis of harvesting costs).

The present value of cash flows

As already shown, there are in theory 30 possible investment alternatives: harvesting from month zero through month 29. Although the fish may be sold in any month, harvesting is relevant only in a shorter period. This analysis will

Table 4.4 Present Value of Cash Flows From Harvesting in July of Year 1 to October Year 2

Year 1

	July	August	September	October	November	December
Discounted sales revenue	477583	643354	888949	1173082	1412122	1486722
Discounted harvesting costs	202205	197200	192318	187558	182914	178387
Discounted feed costs	286049	349032	413884	470442	495084	508306
Present value	−10671	97122	282747	515082	734124	800129

Year 2

	January	February	March	April	May	June
Discounted sales revenue	1520212	1541844	1561635	1583850	1610733	1633353
Discounted harvesting costs	173972	169667	165467	161370	157377	153480
Discounted feed costs	521576	534886	548215	561217	574679	587807
Present value	824664	837291	847953	861263	878677	892066

Year 2

	July	August	September	October
Discounted sales revenue	1655269	1634891	1618537	1600823
Discounted harvesting costs	149681	145977	142265	138842
Discounted feed costs	590328	603857	608318	608318
Present value	916260	885057	867954	853663

evaluate harvesting in the period July year one to October year two, i.e. 16 alternatives are considered. The present values of cash flows from harvesting in these 16 months will be compared to find the alternative that maximizes the present value. For this analysis Table 4.4 is used. It shows the present value of cash flows from harvesting in the respective 16 months.

The discounted sales revenue (biomass value) gives potential sales revenue by months of harvest. These values are taken from Table 4.2.

Harvesting costs are of relevance only if harvesting actually takes place. If harvest occurs in July of year one, there are 75 727 fish (see Table 4.1). With a harvesting cost of $3.10 per fish, nominal harvesting costs equal $234 753.70. Discounted back to the time the fish are released, this is $202 205.00. Observe that harvesting costs diminish over time. This is because the number of fish is reduced due to natural mortality and the present value diminishes due to discounting. By postponing harvesting, these costs are reduced (see Chapter 5).

In the table the discounted feed costs are shown. These are taken from Table 4.3 and increase steadily over time.

The present value of harvesting in the respective months is shown in Table 4.4. This takes into account the fact that harvesting costs are only incurred in the month of harvesting while feed costs are incurred during the entire lifetime of the fish. Table 4.4. shows that the present value is maximized when harvesting in July of year two. This is the same month as the month of maximum discounted biomass value (Figure 4.2).

The present value in July is $916 260. If the fish were harvested one month earlier, the present value would be $24 194 lower, whereas harvesting in August would imply a loss of $31 203. This is a loss of 2.6–3.4 per cent of the maximum present value. The question of optimal harvesting time will be discussed further. A change in the parameter values might imply a change in the harvesting plan. Due to the considerable amount of calculations involved, however, no sensitivity analysis will be undertaken.

Selective harvesting

With the assumptions underlying this analysis, it is optimal for the fish farmer to harvest all fish in the same month. With other assumptions it might be optimal to spread the harvesting over time. This can be due to differences in growth (see Chapter 3) or seasonal price variations. Lower harvesting costs can be envisioned if harvesting is spread over time as this work could be undertaken to a larger degree by the normal labour force. In Figure 4.3 the present values of cash flows from harvesting in one and four months respectively are shown. The first alternative is the same as the one in Table 4.4. In the second alternative, about 25 per cent of the biomass weight is harvested per month over four months.[5] The price and cost parameters are the same.

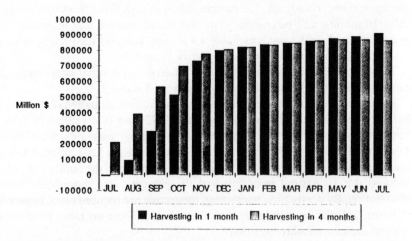

Fig. 4.3 The present value of cash flows of (1) harvesting in one month and (2) harvesting in four months for the period July of Year one to July of Year Two

The figure illustrates that harvesting over four months gives a slightly lower *maximum* present value than harvesting all the fish in one month. The month of maximum present value is June of year two under selective harvesting compared to July of year two when all fish are harvested in the same month. Interestingly, the difference between these two alternatives is small. This result is in fact not surprising, as minor deviations from the optimal harvesting time were shown to imply only a small reduction in the present value. The practical consequences of this result is that harvesting spread over a somewhat longer time period is an alternative to harvesting in one month only.

There are several reasons why selective harvesting can be optimal. As mentioned, this is the case when fish growth differs. Another biological reason for early harvesting would be premature sexual maturation. Spreading the harvest over a longer period might enable the farmer to undertake all harvesting with the normal labour force, whereas hiring additional labour is required when harvesting in a short period. Reduced costs by spreading harvesting over time may by itself make this an optimal policy.

A similar effect may occur on the price side. Many farmers harvesting at the same time, i.e. when the biomass value reaches its maximum, will provide a substantial supply of fish and thus put pressure on the price. This can be avoided by spreading the harvesting over time. There are also certain times of the year when demand is particularly high. By harvesting in these periods, higher prices may be realized.

Earlier it was shown that due to supply and demand conditions there are

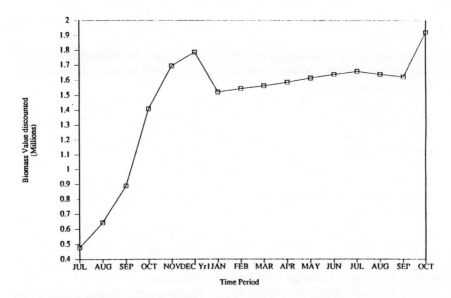

Fig. 4.4 Discounted biomass value from July of Year One to November of Year Two with a 20 per cent price bonus in the months of October, November and December

seasonal variations in price. This can easily be included in the analysis by including a seasonal factor in the price. As an example of this, assume that there is higher demand for salmon before Christmas. Specifically, assume that this implies a price bonus for the fish farmer of 20 per cent when delivering in the months of October, November and December. The discounted biomass value for this case is shown in Figure 4.4.

When compared to the biomass value curve without a price bonus (Figure 4.2), the value curve jumps in the months with the price bonus. Harvesting in these months therefore becomes more attractive than otherwise.

There is no need to elaborate this analysis further. The point has been to illustrate that seasonal variations in price, costs and other factors such as growth rates may imply that it is optimal to spread harvesting over time.

4.3 PRODUCTION PLANNING

This chapter has to a large extent made use of a numerical example to illustrate production planning in a fish farm. To simplify the analysis only one yearclass has been considered. Extending the analysis to include several yearclasses would not introduce any new aspects. Due to the substantial amounts of calculations involved, the consequences of changes in parameter values were not considered.

Personal computer models are a useful tool for such work (see Bjørndal and Uhler, 1989).

This analysis has considered only two types of costs. In practical applications it would be of interest to include more, but that would not alter the qualitative features of our analysis.

The production plan has been evaluated from the time the fish are released. This implicitly assumes that the actual values of the various parameters are identical to the planned ones. This will usually not be the case. Accordingly, such changes must have consequences for the production plan. If, for instance, the growth after one year has been substantially larger than expected, a new analysis for the yearclass must be undertaken in order to determine the optimal harvesting time. This can easily be facilitated within the framework of the model.

The production plan will provide important data for the firm's management system. The relationship between the production plan and the cash-flow budget is obvious. This will be analysed in detail later (Chapter 6). A further objective is to plan production so that harvesting costs are minimized. Moreover, there is an obvious link between production and purchasing.

The kind of analysis undertaken in this chapter is not common in fish farms today. Many of the fish farmers' production decisions will, however, be based on the same kind of evaluations. Increasing competition will necessitate more professional planning and management in fish farms. In this connection, such management tools should have a good future.

NOTES

1 Although harvesting is realistic, for example, during the last 12 months, the farmer may *sell* the fish in earlier months. For the case under consideration, this means there are 30 investment alternatives.

2 In principle revenues in months other than the month of harvesting should also be taken into account. In this example, however, there are sales revenues only in the month of harvesting, so other months can be disregarded.

3 The monthly prices have been calculated on the assumption that the size of the fish is normally distributed with the mean represented by the weights in Table 4.1. Further, based on observations in Norway, it is assumed that the standard deviation is equal to 0.2 of the mean. The *price per kg* has been set to $1.00 for fish less than 1 lb in weight, $4.31 for fish in the 1–2 lb range, $4.87 for the 2–4 lb range, $6.05 for the 4–6 lb range, $7.71 for the 6–9 lb range, and $8.72 for fish weighing more than 9 lb.

4 In Chapter 3 the relationship between the times of maximum biomass value, biomass weight and individual weight is discussed.

5 In the first month, 25 per cent of the fish are harvested, in the second 33 per cent of the remaining fish are harvested, then 50 per cent of the remaining fish

in the third and the remainder of the fish in the final month. Harvesting is assumed to occur at the end of the month, so that fish are fed for the month they are harvested.

5   Cost of production

In this chapter costs in smolt production and fish farming are analysed. Three case studies will be considered: smolt production in Norway, fish farming in Norway and fish farming in British Columbia. The two Norwegian case studies deal with Atlantic salmon, while the one from British Columbia is for Pacific salmon. As far as possible, assumptions that are representative for the industry today are used. Sensitivity analyses will be undertaken in order to investigate how costs vary with certain parameters, and cost of production in Norway and British Columbia will be compared.

In all three case studies monetary values are expressed in *local currencies* (1989 values). At time of printing, the exchange rates are as follows: 1 Norwegian krone (NOK) = 0.18 Canadian $ = 0.15 US $ = 0.09 UK £.[1]

5.1 SMOLT PRODUCTION IN NORWAY

A good location for a hatchery/smolt production unit requires a secure fresh water source (preferably several) of sufficient quality and quantity and it must be situated so that investments in roads, site development and electrical power are manageable. One of the factors determining the magnitude of the investment is the proximity of the facility to the water source. Long water conduits, like long roads, are expensive to build, particularly as there should be several water sources, and accordingly, several conduits in order to reduce risk. As well, the facility ought to be situated near the sea as the ability to take in high quality sea water contributes to a successful smoltification process. Investment will further depend on the necessity of developing new buildings or the possibility of using existing ones. These criteria are applicable to hatchery development anywhere.

Substantial sums are at stake in a smolt operation and the whole year's production may easily be lost. Therefore, it is essential to invest in securing the electric and water supplies. In addition, it has become customary to invest in surveillance equipment for the fish. All this risk reduction has its price.

In Norway, one expects the best locations for smolt operations soon will be exhausted (cf. Section 1.3). This means that a further expansion of smolt production will occur at increased investment costs per smolt.

In the case study under consideration, *all monetary values are expressed in 1989 Norwegian kroner (kr)*.

Facility investments

In the analysis that follows the cost of purchasing the site and water rights is not considered. Neither are very large site investments such as a new access road. Further, it is assumed that the facility must be fully developed before the production begins. It is important to point out that the analysed facility need not be representative, but it has all the essential investment elements.

Investments in a smolt operation (hatchery included) with a production capacity of 1 000 000 Atlantic salmon smolts are as follows:

Site development (planning of lot, water conduits, drainage)	kr	1 500 000 (12.5%)
Facility (building)	kr	4 000 000 (33.3%)
Equipment	kr	5 750 000 (47.9%)·
Interest on construction loan	kr	750 000 (6.3%)
Total fixed investments	kr	12 000 000 (100.0%)

Fixed investments are 12 kr per produced smolt. Interest on the construction loan is calculated as 14 per cent of the average construction loan for one year.

In recent years, financial institutions have been willing to finance fixed investments of 15–20 kr per smolt. Thus the example considered here is reasonable. The upper size limit for smolt producers in Norway today is one million smolts. However, a number of plants are dimensioned for greater production. Extra investments required to increase production will depend on the degree to which the original facility allows for greater production.

Working capital

It is difficult to establish the 'normal' working capital requirement for a hatchery because modes of production vary widely. Costs will also vary substantially from firm to firm. This is particularly true for fixed costs such as maintenance and electricity. Variable costs (roe, wages, feed etc.) will to a greater degree be the same.

The following sets up an *example* of a *production plan* and *operating costs* for a hatchery based on these assumptions:

- The hatchery purchases only eye-roe, not newly hatched roe.
- Price per litre of eye-roe is 2000.00 kr.
- The period from the laying in of eye-roe until hatching is about one month (at a water temperature of 8° C)
- Feeding starts after another month, while the fry will be sold after a further two to three months (at 2–3 g).
- Smolts and fry are sold in the month of May. It is assumed that about 80 per cent of production becomes year-old smolts (this depends *inter alia* on temperature and water quality) and the rest (20 per cent) two-year old smolts
- The conversion ratio is 1.3
- The feed price is 12.00 kr per kg
- For an annual production of 1000000 smolts and 300000 fry, seven workers and a manager are required
- Salaries, including compensation for night duties and social costs, are 200000 kr per man-year, and 250000 kr for the manager
- Vaccination costs 0.70 kr per injection (vaccination)
- Insurance of fish stock: 5 per cent of value
- Insurance of fixed assets: 1 per cent of value
- Other fixed costs include freight, travel, and services

Production plan

An example of a production plan for a smolt producer is:

Dec. 1	Jan. 1	Feb. 1	April 15	May 15 following year
Laying in of eye-roe	→ Hatching	→ Start feeding	→ 3 g fry	→ Smoltification

Mortality

	1.0%	1.0%	15.0%	3.0%

As can be seen, the production period for 'year-old' Atlantic smolts is 18 months. In normal production, without an extraordinary mishap, most of the loss will be in the start-feeding phase (15 per cent) until the fry reach about 3 g. In the period preceding smoltification the loss is assumed to be three per cent (for one-year smolts). An additional one per cent loss is assumed if smoltification occurs after two years.

The mortality is assumed to be evenly distributed over the period. The mortality rates here may appear to be somewhat low. The influence of mortality on production costs will be considered more closely later.

It is assumed that 400 l of eye-roe are laid in yearly at the beginning of December. On average, there will be about 4000 eggs per litre of eye-roe. This volume will permit the sale of fry.

On this basis, the following production plan is arrived at:
May 1st–15th sales:

- 300 000 fry (3 g).
- 800 000 one-year smolts at 50 g.
- 200 000 two-year smolts at 70 g.

In a 'normal' year, there are sales of 1 000 000 smolts and 300 000 fry. Sales of fry very much depend on success in production.

From these assumptions an estimate can be made of the annual operating costs after the facility has come into full operation:

Eye-roe (400 litres)	kr 800 000	(11.8%)
Feed	kr 940 000	(13.8%)
Wages (seven man-years)	kr 1 400 000	(20.6%)
Vaccination	kr 1 420 000	(20.9%)
Insurance fish	kr 500 000	(7.4%)
Total variable costs	kr 5 060 000	(74.4%)
Insurance facility	kr 120 000	(1.8%)
Electricity	kr 500 000	(7.4%)
Maintenance	kr 200 000	(2.9%)
Wage manager	kr 250 000	(3.7%)
Administration	kr 100 000	(1.5%)
Other fixed costs	kr 300 000	(4.4%)
Unforeseen costs	kr 270 000	(4.0%)
Total fixed costs	kr 1 740 000	(25.6%)
Total operating costs	kr 6 800 000	(100.0%)

With an annual production of 1 000 000 smolts, operating costs per smolt become 6.80 kr. This is in the normal range of 5 – 10 kr. Variable costs make up the largest part, 5.06 kr per finished smolt (74 per cent), while fixed costs are 1.74 kr per smolt (26 per cent).

All fry are vaccinated in the fall (about 1 000 000), and all smolts are vaccinated prior to delivery (1 000 000).

Production costs of smolt

Based on the production plan outlined, it is possible to estimate production costs per smolt. However, a change in any of the underlying assumptions will lead to a change in production costs. As it would take too long to analyse the effects of changes in all parameters, the focus will be on *mortality rates* and *capital costs*.

With regard to capital costs, calculations are based on the following assumptions:

- Building: 20 year lifespan
- Special equipment: five year average lifespan
- Site investments: 50 year lifespan
- Real rate of interest: 7 per cent

Annual interest and depreciation charges then become:

Building	kr 400 000.00
Special equipment	kr 1 500 000.00
Site investments	kr 115 000.00
Sum	kr 2 015 000.00

With these assumptions, the following cost breakdown per produced smolt is obtained:

Variable operating costs	kr 5.06	(55.0%)
Interest on working capital	kr 0.40	(4.3%)
Variable costs	kr 5.46	(59.3%)
Fixed operating costs	kr 1.72	(18.7%)
Interest and depreciation on fixed investments	kr 2.02	(22.0%)
Production costs per smolt	kr 9.20	(100.0%)

Interest on working capital is included because of the long production time. Interest is calculated on average capital binding in variable and fixed operating costs for 1.5 years, which is the production time for 'one-year old' smolts. The real rate of interest is set at 7 per cent for both fixed investments and working capital.

All capital costs have been allocated to smolt production, although with the assumed mortality rates some fry will also be sold. This aspect will not be analysed further.

Mortality

In the calculations above a combined mortality from installation of eye-roe to smoltification was set at about 20 per cent in a normal year. Although this is not unreasonable for efficient smolt producers, it is lower than the industry average. Therefore, the sensitivity of production costs to changes in the mortality rate will be analysed. The following alternatives will be investigated:

(1) Estimated mortality from eye-roe to smoltification of 45 per cent.
(2) Loss of 200 000 fry without an opportunity to buy replacement fry.
(3) Loss of 200 000 almost finished smolts.

In the first alternative, the loss is expected beforehand, and extra eye-roe is purchased initially. In the other two alternatives, the loss cannot be replaced. The production costs are given in Table 5.1.

It can be noted that even a strong increase in *expected mortality* leads to only a modest increase in average production cost per smolt. Expected mortality can be compensated for by increasing roe purchases; the cost of roe represents a relatively small part of the production costs per smolt. There is reason to expect that the price of roe will go down with time as production increases. As a result, expected loss will be even less of an economic problem for smolt producers.

On the other hand, *unexpected loss* can lead to a strong increase in average production cost per smolt. The previous examples indicate that a loss of 20 per cent of the fry or smolt will bring about a 22–30 per cent increase in the cost of production. The results also show that the time at which unexpected loss occurs

Table 5.1 Production Costs: Sensitivity to Mortality/Loss Rates and Fixed Investments

	Expected mortality of 20%	Expected mortality of 45%	Unforeseen loss of 200 000 fry	Unforeseen loss of 200 000 smolts	Facility investments 18.2 million kr. Expected morality of 20%
Variable operating costs	kr 5.06 (55.0%)	kr 5.41 (56.5%)	kr 5.63 (50.0%)	kr 6.35 (52.9%)	kr 5.06 (48.3%)
Interest on working capital	kr 0.40 (4.3%)	kr 0.42 (4.4%)	kr 0.45 (4.0%)	kr 0.47 (3.9%)	kr 0.40 (3.8%)
Variable costs	kr 5.46 (59.3%)	kr 5.83 (60.9%)	kr 6.08 (54.0%)	kr 6.82 (56.8%)	kr 5.46 (52.1%)
Fixed operating costs	kr 1.72 (18.7%)	kr 1.72 (18.0%)	kr 2.15 (19.1%)	kr 2.15 (17.9%)	kr 1.72 (16.4%)
Interest and depreciation on fixed investments	kr 2.02 (22.0%)	kr 2.02 (21.1%)	kr 3.03 (26.9%)	kr 3.03 (25.3%)	kr 3.30 (31.5%)
Production costs per smolt	kr 9.20 (100.0%)	kr 9.57 (100.0%)	kr 11.26 (100.0%)	kr 12.00 (100.0%)	kr 10.48 (100.0%)

has little influence on the average production cost. If the loss occurs with 3 g fry, instead of finished smolts, feed, vaccination and insurance costs are reduced. There is little one can do with the other costs in the short term. However, the producer's loss will depend on *insurance arrangements*. Insurance will cover most of the loss if it occurs with finished smolts. For fry loss, the insurance will only cover the value at the time of the loss and not the future value.

Investment costs

The fixed investments stipulated above may appear somewhat low when compared to smolt operations more recently developed. To determine the influence of capital costs on production costs, another alternative for a producer with a capacity of one million smolts is considered. The investments will be as follows:

Site construction	kr 3 000 000	(16.5%)
Facility (building)	kr 4 000 000	(22.0%)
Equipment	kr 10 000 000	(54.9%)
Construction loan interest	kr 1 200 000	(6.6%)
Total fixed investments	kr 18 200 000	(100.0%)

In this alternative, fixed investments are 18.20 kr per smolt. This is not uncommon. This alternative is based on high investments for both site construction and equipment, including surveillance equipment. The production costs are given in Table 5.1, assuming the same level of operating expenses and mortality as in the original case.

Compared to the original case, the result clearly shows that the *level of fixed investments* influences production costs per smolt. The consequence of this will be seen as pressure on the smolt price increases due to the substantial increase in smolt supply caused by expanded production capacity (cf. Section 1.3).

If the facility is dimensioned to produce more than 1 000 000 smolts, an expansion of production capacity would lead to a reduction in average production costs. This would be an indication of economies of scale in smolt production. There is reason to believe that these are not insignificant as capital costs comprise a relatively large portion of total production costs.

In these calculations all costs are allocated to smolt production, even though the operation also produces some fry. This is because it is difficult to correctly allocate the costs between the two products. Nonetheless, it should be pointed out that if the firm also sells fry, it will be possible to break even with prices lower than those calculated.

5.2 SALMON FARMING IN NORWAY

In this section the development of a fish farm with a pen-volume of 8000 m³ producing Atlantic salmon will be analysed. An estimate of the facility investment and associated working capital requirements will be given. Production costs are analysed.

All monetary values are given in Norwegian kroner (1989 values).

Facility investments

Fish farming is undertaken in different modes. Traditionally a sea-pen system has been attached to a land-based facility. Investments can then be divided into *land facilities* (road, building, pier, etc.) and the *sea-pen system* itself. Experience shows that investments on land will vary greatly, depending on existing facilities. Many farms are established in connection with existing fish-processing plants where one has piers, freezers and buildings. Other farms may be started from scratch and require access roads, piers and a building for storage and processing. The type of building depends on the type of feed used. Moist feed requires a freezing room, sour feed requires tanks and dry feed requires a dry storage room. Moreover, whether one slaughters and packages the fish oneself will influence investment costs. It is therefore difficult to find a general definition of the 'common' investments on land.

On the other hand, investment in the sea-pen system will to a large degree be the same. Today, a newly established farm is licensed for 12 000 m³. Nevertheless, initial capacity will vary as it takes two to three years from when the first smolt is set out until the farm can reach full production. All the same, many investors dimension the farm for 12 000 m³ right away. With regard to feeding, choice of technological standard will be the deciding factor for the investment need. The larger the degree of automation, the larger the investment. A certain degree of automation of feeding routines is today common in many farms.

Pollution has become a problem for many farmers and the productivity of a location is not fully known when a new farm is established. For both these reasons, flexibility with respect to location is advantageous. As a consequence, in the last few years many farms have been developed as *floating units*. A float contains all required facilities with the sea-pen system attached to the float. This makes it possible to move the whole farm from one location to another. For these reasons, development of a floating farm will be considered here. A farm with 8000 m³ pen capacity will be analysed, as this is the 'typical' size for Norwegian producers today. Enlargement to 12 000 m³ will also be considered.

The following overview of necessary investments in a new 8000 m³ farm is based on information from supply manufacturers:

Float	kr 825 000	
House	kr 270 000	
Feeding equipment with silo	kr 600 000	
Generator, watertank, hydrofor (installed)	kr 150 000	
Gangway	kr 100 000	
Anchoring	kr 200 000	
Float with equipment		kr 2 145 000
Pens (10 pens)	kr 1 180 000[2]	
Transportation, installation and anchoring	kr 200 000	
Nets (30 ft × 10)	kr 300 000[3]	
Safety equipment	kr 120 000	
Cranes/winches	kr 50 000	
Sea-pen system		kr 1 850 000
Boats		kr 175 000
Miscellaneous		kr 80 000
Total investments		kr 4 250 000

Total investments are 530 kr. per m^3. This is considered 'normal' for the industry. It is assumed that the farm is a 'turn-key' installation; therefore interest on construction capital is irrelevant.

The float will accommodate enlargement to 12 000 m^3. Most additional investments are in the sea (extra pens, nets and so forth). This case will be considered below.

Up to a certain size there are advantages to expanding the pen volume, depending on the capacity of the fixed investments. The point is to use capacity (building, storage, processing, etc.) as efficiently as possible. This can be done either by increasing own volume or through cooperation with other farmers. The investment per farm can be considerably reduced if several farmers cooperate in feed production, fish slaughter, equipment and smolt purchase and so forth.

Working capital

The following farm *production plan* and estimated operating costs are based on the following assumptions:

- 80 000 salmon smolts are set out every year in May
- Price per smolt is 13 kr
- Feed price is 8 kr per kg
- Conversion ratio = 1.3

- Mortality: The first month after release – seven per cent
    The next six months – four per cent
    Further – two per cent per half year
- Insurance fish: 1.50 kr per kg produced
- Insurance facility: one per cent of value
- Labour: Three workers and a manager
- Wages including social costs 200 000 kr per worker and 250 000 kr for farm manager
- Harvesting begins about 15 months after smolt release and continues at the same rate over the year
- Average weight per fish is 3.7 kg at harvest

With these assumptions about 68 000 fish with an average weight of 3.7 kg will be produced for a total production of 250 t per year. The following budget is for annual operating expenses:

Variable:

Smolts	kr 1 040 000 (18.6%)
Feed	kr 2 670 000 (47.8%)
Wages	kr 600 000 (10.7%)
Insurance fish	kr 385 000 (6.9%)
Total	kr 4 695 000 (84.0%)

Fixed:

Wage manager	kr 250 000 (4.5%)
Administration	kr 100 000 (1.8%)
Electricity	kr 100 000 (1.8%)
Insurance facility	kr 40 000 (0.7%)
Maintenance	kr 200 000 (3.6%)
Travel	kr 100 000 (1.8%)
Miscellaneous	kr 100 000 (1.8%)
Total fixed costs	kr 890 000 (16.0%)
Total operating costs	kr 5 585 000 (100.0%)

In a 'normal' year operating expenses are 5 585 000 kr. With a production of 250 t of salmon per year, operating costs are 22.34 kr per kg of produced fish. Most of these are variable costs (18.78 kr or 84.0 per cent), while the fixed costs amount to 3.56 kr per kg (16.0 per cent).

Production costs for farmed salmon

Many variables influence production costs for farmed salmon. Considered here are *smolt prices*, *mortality rates* and *conversion ratios*. Interest on working capital and interest plus depreciation on the invested capital must also be considered. All this is done for a 'normal' year. As well, the consequences of expanding the farm size are analysed. These estimates will be compared to the results from a cost study undertaken by the Directorate of Fisheries.

It is difficult to determine the lifespan of the fixed investments. Estimates are based on the following assumptions:

- Feeding equipment   5 years
- Nets   3 years
- Other investments   10 years
- Real rate of interest   7 %

With these assumptions, annual interest and depreciation charges become:

Feeding equipment	kr 145 000.00
Nets	kr 115 000.00
Other investments	kr 475 000.00
Sum	kr 735 000.00

Production costs per kg are given as alternative A in Table 5.2. Interest on working capital is calculated on average capital binding for variable and fixed costs for two years.

Feed comprises the largest share of costs. One kg of fish produced requires 1.3 kg of feed. In addition, feed is used for fish that later die. Taking that into account, total feed cost per kg of production is 10.70 kr. Reducing feed loss will improve profitability.

Changes in the production plan will also cause changes in production costs per kg. Three parameters are considered: *smolt price*, *mortality rate* and *conversion ratio*. The following alternatives are assessed:

(A) The basis-alternative above.
(B) Conversion ratio of 1.7; otherwise as (A).
(C) Smolt price of 9 kr; otherwise as (A).

All cases consider *expected mortality*. With the assumptions regarding fish release and mortality, all three cases comprise a yearly production of 250 t. This entails the same degree of utilization for all cases, so they can be compared directly. The results are given in Table 5.2.

Of total production costs per kg of salmon for the different alternatives, smolts account for 10–15 per cent, feed 40–45 per cent, labour 10 per cent, and interest on working capital about 5 per cent. Fixed costs, including depreciation, maintenance and a normal return on capital, are about 25 per cent. Feed is

Table 5.2 Average Production Costs per kg Fish

	(A) Smolt price = 13. kr Mortality = 15% Conversion ratio = 1.3 Production = 250 t	(B) Smolt price = 13. kr Mortality = 15% Conversion ratio = 1.7 Production = 250t	(C) Smolt price = 9. kr Mortality = 15% Feed factor = 1.3 Production = 250 t	(D) Smolt price = 9. kr Mortality = 15% Feed factor = 1.3 Production = 375 t
Smolt	kr 4.16	kr 4.16	kr 2.88	kr 2.88
Feed	kr 10.70	kr 14.00	kr 10.70	kr 10.70
Wages	kr 2.40	kr 2.40	kr 2.40	kr 2.00
Insurance fish	kr 1.50	kr 1.50	kr 1.50	kr 1.50
Interest working capital	kr 1.34	kr 1.46	kr 1.16	kr 1.11
Variable costs	kr 20.10	kr 23.52	kr 18.64	kr 18.19
Fixed costs	kr 3.56	kr 3.56	kr 3.56	kr 2.96
Interest and depreciation on fixed investments	kr 2.94	kr 2.94	kr 2.94	kr 2.41
Production costs per kg	kr 26.60	kr 30.02	kr 25.14	kr 23.56

clearly the most significant input cost in salmon production.

A conversion ratio of 1.7 (alternative B) as compared to 1.3 (alternative A) involves higher variable costs of 3.42 kr per kg or 17 per cent. This difference shows a potential for cost saving as feed costs comprise a considerable part of total costs. A rise in the feed price will have the same impact.

A reduction in the smolt price from 13 kr to 9 kr (alternative C) has been considered because the increase in smolt supply has caused a price fall. This leads to a reduction in variable costs of 1.46 kr per kg or 7 per cent. Thus a relatively large change in the smolt price will have only a moderate effect on production costs.

Finally, production cost estimates are made based on the facility being expanded from 8000 m³ to 12 000 m³. This will provide an increase in production from 250 t to 375 t per year. Assume that due to the expansion, the following additional investments are required:

Extra pens (five)	kr 500 000.00
Feeders	kr 90 000.00
Freight and installation	kr 50 000.00
Nets	kr 150 000.00
Safety equipment	kr 60 000.00
Sum additional investments	kr 850 000.00

Assuming that a 25 per cent increase in the fixed operating and the wage costs is required, the result is given as alternative (D) in Table 5.2.

Under these assumptions, the production costs are reduced by 1.58 kr per kg compared to alternative (C). Cost reduction takes place through lower capital and fixed costs per kg produced plus lower labour costs due to more efficient use of labour. In addition, capital binding in production becomes lower and this leads to less interest on working capital.

In general, lower production costs can be achieved through:

(1) reduced prices on production factors, such as smolt and feed;
(2) more efficient use of inputs such as capital equipment, labour and feed (reduced conversion ratio);
(3) more efficient plant size.

While the fish farmer has little influence on prices of the factors of production, he can have some impact on the use of feed, capital equipment and labour. In regard to point three above, the reduction comes about by expanding the facility which may lead to more efficient use of the equipment.

In investigating economies of scale in the industry, Salvanes (1988) analysed the costs of fish farms in 1982 and 1983. He found that average production costs per kg decrease when the production quantity goes up; in other words, there are

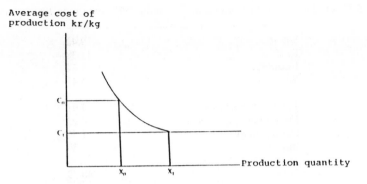

Fig. 5.1 Average cost of production per kg

economies of scale in the production of farmed salmon. This is illustrated in Figure 5.1.

The figure shows a declining average cost curve. This is in accordance with Salvanes' results and indicates economies of scale in production. This could be due to more efficient use of inputs such as capital and labour as well as more efficient farm size. Nevertheless, there will be an upper limit on efficient farm size. Too many fish in one location can lead to more pollution, disease, a higher degree of stress for the fish and, consequently, higher production costs per kg. This is indicated by the flattening out of the average cost curve for a production quantity of X_1 in Figure 5.1. In the case of the data-set employed by Salvanes, this flattening occurred at an output level of about 110 t. This is less than the output assumed for the 'standard' 8000 m³ farms (Table 5.2). The reason for this is that Salvanes' data-set is from 1982–1983, with most farms not yet fully developed. With newer data this flattening would take place at a higher output level.

Another result worth noting is that the difference in average production costs per kg between output levels of 50 and 110 t is only about 10 per cent. This indicates that while economies of scale in production do exist, they are modest.

The *location* of the farm has much to do with natural productivity and, therefore, production costs. In this way, an optimum size in one location may be 5000 m³ and 30 000 m³ in another. Work is under way to discover how and to what degree location affects production costs.

Since 1982 the Norwegian Directorate of Fisheries has undertaken annual cost and profitability studies for fish farms. The last available report is from 1988. Its main results are given in Table 5.3. The Directorate of Fisheries' results cannot be compared directly with the results from this analysis because their sample included many farms that were not fully developed. This can be seen from the smolt costs (9.42 kr per kg of farmed salmon, see Table 5.3). Smolts purchased in 1988 would be harvested one or two years later but, in the

Table 5.3   Production Cost per kg Farmed Salmon, 1988. Figures in 1988 kroner

Smolts	9.42
Feed	11.38
Insurance	1.64
Wages	5.47
Total variable costs	27.91
Fixed operating costs	4.93
Depreciation	1.56
Interest[a]	6.39
Total	40.79
Average production quantity (tonnes)	188.70
Workers per farm	2.90
Sample size	281

[a] Interest includes interest on debt and calculated interest on equity.

Source: Directorate of Fisheries, Bergen, Norway: Profitability Study of Fish Farms, 1988.

calculations of the Directorate of Fisheries, were accounted for in the 1987 costs. For farms that are in the process of expanding production, as is the case for most farms, this method of calculation leads to an overestimate of costs. The same happens with regard to fixed costs, including labour costs in so far as labour costs are fixed. To the degree that production capacity is not fully exploited, average costs per kg are overstated.

The analysis undertaken in this chapter is based on full capacity utilization. Seen in this way, the farms represent 'model' facilities. In reality, it may take time for farms to reach the efficiency level modelled here. This is in part due to the lack of proper management. However, the results indicate that there is substantial scope for further cost savings.

5.3 SALMON FARMING IN BRITISH COLUMBIA

British Columbia is expected to become a major producer of farmed salmon in the 1990s. The dominant species being farmed are chinook and coho, but Atlantic salmon has also been introduced. However, due to restrictions on the importation of Atlantic eggs and the long production cycle for this species, it will be some time before the production of Atlantic salmon becomes important.

Most farms in British Columbia today produce both chinook and coho salmon. To a large extent this is due to a limited availability of chinook smolts. Certainly the production of chinook salmon appears to be more profitable than the production of coho and as the shortage of chinook smolts is alleviated, it can be expected that they will represent a larger proportion of total output. However, it should be noted that the lifecycle of coho is shorter than that of chinook (cf. Section 1.1) so that coho can generate a positive cash-flow faster than chinook. If financing is a constraint on farm development and construction, this represents a reason for also producing coho.

In this section, cost of production for a *model farm* in British Columbia will be analysed. As it is assumed that financing represents no constraint on development, a farm that produces only chinook will be considered. The analysis will be for a fish farm with an annual production capacity of 400 t. This may represent average plant size in the industry, although industry structure is not yet settled in British Columbia.

All figures are expressed in *Canadian dollars* (1989-values).

Facility investments

The infrastructure in British Columbia is less well developed than in Norway. Many areas with good biophysical conditions for farming are inaccessible by road and have no electric power supply. Thus, transportation will be by sea and electric generators will have to be built. Investments in land facilities per unit of production capacity may therefore be higher than in Norway. For these reasons, many farms in British Columbia have been developed as floating units.

Although investments in the *sea-pen system* vary less from farm to farm, they are also higher per unit of production capacity than in Norway because of lower stock densities. While stock densities for Atlantic salmon typically are 20–25 kg/m^3, the maximum for Pacific salmon is generally set at 8 kg/m^3, with lower densities not uncommon. As the Pacific salmon species become more domesticated, this will result in higher permissible stock densities. However, due to the long regeneration period for salmon, this may be a lengthy process.

Unlike Norway, farm size is not regulated in British Columbia. This will allow British Columbia farmers to utilize such economies of scale as exist in production. Allowing farmers to operate at the most efficient output level may result in cost advantages vis-à-vis Norwegian producers. Furthermore, the industry may exploit cost savings attributable to horizontal and vertical integration, something Norwegian regulations deny growers there.

In the present case study, a floating farm will be considered. This will facilitate direct comparison with the Norwegian case. The following development and equipment investments will be used:

Float with house	$120 000
Feeding equipment with silo	$140 000
Generator, watertank, hydrofor (installed)	$ 30 000

Gangway	$ 20 000	
Anchoring	$ 40 000	
Float with equipment		$350 000
Pens (20 pens)	$430 000[4]	
Transportation, installation and anchoring	$ 40 000	
Nets (30′ × 10)	$110 000	
Cranes/winches	$ 10 000	
Sea pen system		$590 000
Boats		$ 30 000
Miscellaneous		$ 10 000
Total investments		$980 000

Compared to the Norwegian farm, the investment in a float with house is less in British Columbia. However, due to lower stocking densities, investments in pens are relatively higher in BC. Safety equipment has been disregarded in BC, as this is not required.

Working capital

The following *production plan* and estimated operating costs for the chinook salmon farm are based on the following assumptions:

- 180 000 salmon smolts are set out every year in April
- Price per smolt is $0.85
- Feed price is $1.30/kg
- Conversion ratio = 1.7
- Natural mortality = 1.5 per cent per month
- Labour: four workers and a manager
- Wages: $25 000 per year for workers and $35 000 for farm-manager (including social costs)
- Fixed costs include equipment insurance, electricity, overhead, maintenance and medicine
- Insurance of fish is calculated on the basis of the sum of smolt and feed costs at a rate of 4 per cent per annum
- Harvesting begins about 19 months after release of smolts and continues at the same rate over the year
- Average fish weight is 3.06 kg at harvest

Compared to the Norwegian case study, some differences can be observed. For

Atlantic salmon there is a fairly high initial mortality after the release of smolts, for chinook the mortality is spread more evenly over time. However, total mortality from time of release to harvesting is higher for chinook, presumably due to a lower degree of domestication. By itself, this would contribute to higher unit production costs for chinook. On the other hand, the smolt price is lower for chinook than for Atlantic salmon. This lowers production costs. Another difference is that the average weight at harvest is lower for chinook than for Atlantic salmon.

With these assumptions, harvesting will start in the fall of the second year of operation. Normal production will be reached in year three, when 131 000 fish averaging 3.06 kg will be harvested. This generates a total production of 400 t per year. The combined mortality is 27 per cent, compared to 15 per cent in the base case for the Norwegian farm.

The following budget is for operating expenses:

Variable	
Smolts	$ 153 000
Feed	$ 850 000
Wages	$ 135 000
Fish insurance	$ 40 000
Sum variables	$1 178 000 (85.5%)
Miscellaneous	
fixed costs	$ 200 000 (14.5%)
Total costs	$1 378.000 (100.0%)

In a normal year operating expenses are $1 378 000. With a production of 400 t of salmon per year, operating costs are $3.45 per kg of produced salmon. Most of these are variable costs ($2.95 or 85.5 per cent), while the fixed costs amount to $0.50 per kg (14.5 per cent). It is noteworthy that unit operating costs are somewhat lower for the British Columbia producer than for the Norwegian.

Interest and depreciation must be taken into account when calculating production costs for farmed chinook salmon. As in the previous case studies, this is done for a 'normal' year.

The following assumptions are made regarding the lifespans of the fixed investments:

● Feeding equipment	5 years
● Nets	3 years
● Other investments	10 years
● Real rate of interest	7 %

As will be recalled, these are the same assumptions as in the Norwegian case.

Table 5.4   Average Production Costs per kg of Chinook Salmon

Smolt	$0.38
Wages	$0.34
Feed	$2.13
Fish insurance	$0.10
Interest on working capital	$0.20
Variable Costs	$3.15
Miscellaneous fixed costs	$0.50
Interest and depreciation on fixed investments	$0.45
Production costs per kg	$4.10

This will facilitate a comparison between the two cases. With these assumptions, annual interest and deprecation charges become:

Feeding equipment	$ 42 000
Nets	$ 34 000
Other equipment	$104 000
Sum	$180 000

Table 5.4 summarizes average production costs for a 'normal' year.

Under the assumptions made for the 400 t chinook farm, average production costs are $4.10 per kg. As anticipated, feed is the largest cost component, representing 52.0 per cent of total costs.

It is of interest to compare the production costs for chinook and Atlantic salmon. At current exchange rates, the smolt cost is much higher for Atlantic salmon (cf. Table 5.2). As the price of Atlantic smolts is reduced over time, this difference will become less. However, as the production time for Pacific smolts is considerably less than for Atlantic smolts, Pacific smolts will remain cheaper.

Feed costs are somewhat higher for Pacific salmon. As the feed prices are comparable, the difference can be attributed to a higher conversion ratio in BC and a higher mortality for chinook salmon. If the mortality rate and conversion ratio for chinook can be brought down to the level of Atlantic salmon, this cost difference will disappear.

Wages are lower in British Columbia, primarily due to lower wage rates than in Norway. Interest on working capital is also less, as are total variable costs. Interest and depreciation on fixed investment are somewhat higher in British Columbia. This is due to lower stock densities for chinook salmon. Although farmers in British Columbia may utilize possible economies of scale in production and enjoy savings due to horizontal and vertical integration, these factors are not sufficient to outweigh the effect of lower stock densities.

A word of caution is in order at this point. The Norwegian industry is more established than British Columbia's; the longer operating history makes for more accurate production costs. Thus, cost estimates are more reliable for Norway than for British Columbia. However, the potential for cost savings due to improved husbandry and economic rationalization is substantial in both countries.

Salmon farming is not yet well established in most countries and so far, relatively few analyses of cost of production have been undertaken. Those that do exist tend to be engineering analyses rather than based on farm-level micro-data.[5] Cost of production in salmon aquaculture is therefore an interesting area for further research.

NOTES

1 These are exchange rates in January 1990. Source: Den norske Creditbank.
2 Dimension: 15 m × 15 m.
3 It is assumed that two nets are purchased for each pen.
4 Dimension 15 m × 15 m.
5 See Shaw and Muir (1987) and DPA (1988).

6 Start up of a fish farm

So far, operations of a farm in full production have been analysed (Chapters 3–5). Now problems related to the development of a new farm will be considered.

The production period in salmon aquaculture is long. During the development period, which may last for several years, there will be little or no sales revenue. In this period all expenditures must be financed out of equity or loan capital. Thus financial planning is very important. For this reason special emphasis will be placed on cash-flow budgeting for the first few years of operations.

The investment plan gives information about investments and future financial payments. The production plan provides information about present and future costs and revenue. In previous chapters optimal harvesting and cost of production have been analysed. By combining these with a financial plan and assumptions about the development period, a cash-flow budget can be derived.

The point of departure will be the Norwegian fish farm considered in Section 5.2. First the fixed investments will be considered. Then the production plan will be specified, giving information about costs and revenue over time. With assumptions about financing and inflation, a cash-flow budget for the first five years of operations can be established.

Facility investments

Investments in a fish farm can usually be divided into *land facilities* and a *sea-pen system* and can normally be spread over time. Sea-pen capacity is added gradually as the fish grow and additional yearclasses are released. Nor is there an immediate need for all land facilities and other fixed investments; the development of a farm is usually spread over a two to three year period. This eases financial requirements.

The cases considered in Chapter 5 were *floating units*. These are turn-key installations. When such a technology is chosen, all investments in fixed capital are undertaken up-front prior to the release of the first yearclass of smolts. For the Norwegian case study, fixed investments amount to 4 250 000 kr. These are undertaken in year one.

Production Plan

It is assumed that 80 000 salmon smolts are set out every year in May. The first yearclass is released in May of year one. Harvesting starts after 15 months, i.e. in August of year two and continues until May of year three. The release and harvest pattern is the same for subsequent yearclasses. (The second yearclass is released in May of year two, the third in May of year three and so forth.) Average weight at harvest is 3.7 kg. (See Section 5.2 for assumptions about natural mortality.) The following production plan is arrived at:

Half year period	Harvest from yearclass (tonnes)				Harvest	
	1	2	3	....	per period	per year
1st–3rd	—	—	—			
4th	69	—	—		69	69
5th	181	—	—		181	
6th	—	69	—		69	250
7th	—	181	—		181	
8th	—	—	69		69	250
9th	—	—	181		181	
.						
.						
.						
Sum	250	250	250			

The first half year refers to the period January – June when the first yearclass of smolts is released (year one).

Each yearclass yields an output of 250 t. Harvesting starts in year two with 69 t from the first yearclass. In year three one harvests both from the first yearclass (181 t) and the second (69 t), in total 250 t. The farm is in full production from year three onwards with an annual output of 250 t. This is equal to total yield from one yearclass.

Working capital

In the first year of operations, it is assumed that two workers are required. Full production requires four workers. Based on the production plan and other assumptions regarding inputs and input prices, the following budget for operating expenses is derived:

Operating expenses ('000 1989 kr)

	1st half year	2nd half year	3rd half year	4th half year	3rd year and later
● Variable					
Smolts	1040		1040		1040
Feed	80	450	850	1580	2670
Insurance fish	10	65	125	225	385
Wages	100	200	300	300	600
Total variable costs	1230	715	2315	2105	4695
● Fixed:					
Wage manager	125	125	125	125	250
Administration	50	50	50	50	100
Travels etc.	50	50	50	50	100
Electricity	25	40	50	50	100
Insurance facility	25	50	75	75	200
Maintenance	25	50	75	75	200
Miscellaneous	100	75	50	50	100
Total fixed costs	400	440	475	475	1050
Total operating costs	1630	1155	2790	2580	5745

Operating expenses increase over time as production is expanded. Smolts are purchased in May. Therefore operating expenses are higher in the first six months of the year.

When the farm reaches normal production (year three or later) annual operating expenses are 5745000 kr. Most of these are variable costs.

Cash-flow budget

To set up a cash-flow budget, the following assumptions about prices, financial payments and conditions of payment are made:

● Average price per kg of salmon is 35 kr (round weight)
● All sales revenue is received in the year of sale
● All payments occur per 30 days, with the exception of wages
● All facility investments are financed by long term loans and equity. (A financial plan is set up below)
● Operating costs are financed by a line of credit. The interest rate on the line of credit is 15 per cent per annum plus a fixed (stand-by) charge of 1.5 per cent of the limit. The interest charges are calculated on the basis of average utilization of the line of credit and, together with the fixed charge, are drawn on the line of credit at the end of every half year.

The facility investments amount to 4 250 000 kr. It is assumed that they are financed as follows:

Trust company loan	kr 2 000 000 (47.1%)
Bank loan	kr 1 250 000 (29.4%)
Equity	kr 1 000 000 (23.5%)
Sum	kr 4 250 000 (100.0%)

In this example, equity represents 23.5 per cent of facility investments. Operating costs are financed by a line of credit. Operating expenses are normally financed in part by equity, so more equity will be required than has been assumed in this example.

We assume the following conditions for the loans:

Trust company
- Interest rate of 15.0 per cent p.a. paid every six months
- No repayments on principal for the first two years; thereafter repayment over 10 years

Bank
- Interest rate of 16.0 per cent p.a. paid quarterly
- Annuity loan over eight years; no repayment on principal the first year

On this basis, financial payments for the first years will be (in '000 Kr):

	1st half year	2nd half year	3rd half year	4th half year	3rd year	4th year	5th year
Interest payments:							
Trust company	150	150	150	150	292.5	262.5	232.5
Bank	100	100	99	95	176	155	129.5
Total interest payments	250	250	249	245	468.5	417.5	362
– Payments on principal	—	—	51	55	324	345	370.5
Interest and principal payments	250	250	300	300	792.5	762.5	732.5

The following assumptions about inflation are made:

- The sale price of salmon increases at an annual rate of 4 per cent
- The price of smolts is constant in nominal terms
- All other prices increase at an annual rate of six per cent
- Inflation is spread evenly over the year

With these assumptions, the real price of both salmon and smolts will decrease. The cash-flow budget (expressed in nominal values) for the farm for its first five operating years is given in Table 6.1. It has been assumed that the limit on the line of credit is increased gradually, so that fixed charges will not be larger than necessary. During the development period for the farm (years one and two), there is a rapid increase in capital requirements as all expenses are paid out of a line of credit.

Harvesting begins in August of the second year of operations, i.e. in the fourth half year period. This is when the capital requirement reaches a maximum of about 8.0 million kr. This represents 32 kr per kg of fish produced. This figure clearly indicates the need for prudent financial planning when developing a farm. With no extraordinary production mishaps, a farm will be able to start repaying its credit in year three. From that year onwards, the farm yields a positive cash flow.

The cash-flow budget is set up on a half yearly basis for the first two years and thereafter on a yearly basis. This represents a somewhat rough periodization.

Table 6.1   Cash-flow Budget for Fish Farm ('000 nominal kr)

	1st half year	2nd half year	3rd half year	4th half year	3rd year	4th year	5th year
Sales revenue	–	–	–	2585	9665	10040	10440
Operating expenses	–1535	–1120	–2775	–2625	–6320	–6590	–6930
Financial payments	– 250	– 250	– 300	– 300	– 795	– 765	– 735
Capital requirement current period prior to interest charges	–1785	–1370	–3075	– 340	2540	2685	2775
Capital requirements previous periods	–	–1865	–3450	–6955	–7890	–6450	–4605
Capital requirement prior to interest charges	–1785	–3235	–6525	–7295	–5350	–3765	–1830
Interest charge, line of credit	– 65	– 190	– 375	– 535	– 990	– 765	– 480
Stand by charge	– 15	– 25	– 55	– 60	– 110	– 75	– 35
Capital requirement	–1865	–3450	–6955	–7890	–6450	–4605	–2345
Change in cash-flow	–	–1585	–3505	– 935	+1440	+1845	+2260
Limit, line of credit	2000	3500	7300	8000	7300	5000	2400

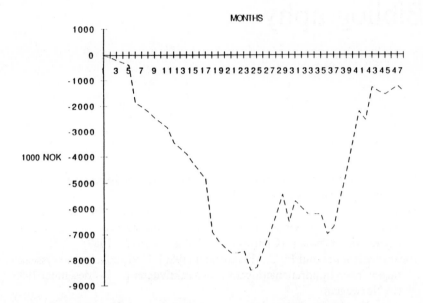

Fig. 6.1 Development in line of credit for newly developed fish farm

For this reason Figure 6.1 shows the development in the line of credit on a monthly basis for the first four years. This development depends on the timing of sales receipts and payments, that is whether sales revenue is received before or after payments of interest charges and other expenses. For this reason, the figure should only be considered as an example.

When comparing the figures in Table 6.1 with Figure 6.1, it can be noted that the cash-flow budget – on a half yearly or a yearly basis – causes a certain averaging out of the capital requirements. This might lead to an underestimation of the capital requirement.

The point to be made here is that *periodization* is important. Choice of periods for the cash-flow budget must be such that one does not specify periods so long that the capital requirement is underestimated. This is even more important for hatcheries (sales once or a few times per year) than for fish farms. In such cases it is important to undertake monthly budgeting.

In this case study, only one example has been considered. When planning start-up of a farm, several alternatives should be analysed with respect to cash-flow. This would include different scenarios as to input and output prices and other variables such as the conversion ratio and growth rates. In addition the effects of extraordinary loss due to unexpected mortality should be analysed.

Bibliography

Allen, P.G., Botsford, L.W., Shuur, A.M. and W.E. Johnston (1984). *Bioeconomics of Aquaculture*. Amsterdam: Elsevier.

Anderson, J.L. (1987). Strategic Design and Marketing of Aquacultured Salmon. Paper presented at the symposium on markets for seafood and aquaculture products (Charleston, SC, August 19–21, 1987).

Arbeidsdirektoratet and Fiskeridirektoratet (1986). Undersøkelser av sysselsettingen innen oppdrettsnæringens primæraktiviteter pr. 1–7 desember 1985. (In Norwegian.)

Arnason, R. (1987). Optimal Feeding Schedule and Harvesting Time in Aquaculture. Unpublished mimeo.

Atkinson, C.E. (1987). The Fisheries and Markets of Japan with Special Reference to Japan. Paper presented at the Fundacion Chile conference on salmon aquaculture (Santiago, March 17–19, 1987).

Atkinson, C. (1989). Japan. *Salmon Market Newsletter*. Vol. 1, N. 3, pp. 10–11.

BIM (A Board Iascaigh Mhara/Irish Sea Fisheries Board) (1986). *The Atlantic Salmon Farming Industry*. Main Report.

Bjørndal, T. (1987). *Fiskeoppdretts-økonomi*. Oslo: J.W. Cappelen. (In Norwegian.)

Bjørndal, T. (1988). The Optimal Harvesting of Farmed Fish. *Marine Resource Economics* 5: 139–159.

Bjørndal, T. and Salvanes, K. (1987). Ramevilkår for næringa (Regulations of the Industry). In *Fiskeoppdretts-økonomi*, ed. T. Bjørndal. (In Norwegian.)

Bjørndal, T. and Schwindt, R. (1987). Norwegian Direct Investment in the British Columbia Salmon Aquaculture Industry: A Case Study. Discussion Paper No. 1/1987, Institute of Fisheries Economics, Norwegian School of Economics and Business Administration.

Bjørndal, T., and Uhler, R.S. (1989). Salmon Sea Farm Management: Basic Economic Concepts and Applications. Working Paper, Centre for Applied Research, Norwegian School of Economics and Business Administration.

Bjørndal, T., Schwindt, R. and Marshall, L. (1989). The Regulation of Salmon Aquaculture in Norway, British Columbia and Alaska. Discussion Paper, Institute of Fisheries Economics, Norwegian School of Economics and Business Administration.

British Columbia, Ministry of Agriculture and Fisheries (1988). 1987 Salmon Farm Survey, Final Report of the Survey Team.

Clark, C.W. (1976). *Mathematical Bioeconomics: The Optimal Management of Renewable Resources*. New York: Wiley Interscience.

Communications Directorate, Department of Fisheries and Oceans (1988). *Commercial Aquaculture in Canada*. Ottawa: Supply and Services.

Department of Agriculture and Fisheries for Scotland: Salmon and Trout Farming in Scotland – Report of DAFS Annual Survey for 1987.

DeVoretz, D.J. and Salvanes, K.G. (1990). The Demand for Pen-Reared Salmon: Market Structure and Stability. Working Paper, Centre for Applied Research, Norwegian School of Economics and Business Administration.

DPA Group Inc. (1986). Industrial Organization of the BC Salmon Aquaculture Industry. Mimeo, Fisheries and Oceans Canada.

DPA Group Inc. (1988). Cost of Production Model for Pen-Rearing of Salmon in Alaska and currently Producing Regions. Reports prepared for State of Alaska, Department of Commerce Economic Development.

Grønhaug, K. (1985). *Markeder og strukturer for oppdrettslaks (Markets and Structures for Farmed Salmon)*. Bergen. Centre for Applied Research. (In Norwegian.)

Helgason, A. (1986). *Production in Icelandic Fish Farming in 1986*. Reykjavik: Institute of Freshwater Fisheries.

Helgason, A. (1987). *Production in Icelandic Fish Farming in 1986*. Reykjavik: Institute of Freshwater Fisheries.

Herrmann, M. and Lin, B.-H. (1988). An Econometric Analysis of the Demand and Supply of Norwegian Atlantic Salmon in the United States and the European Community. Unpublished mimeo (February 22, 1988).

Jóhansson, V. (1988). *Production in Icelandic Fish Farming in 1987*. Reykjavik: Institute of Freshwater Fisheries.

Leitz, P. (1989). Canada. *Salmon Market Newsletter*. Vol. 1, No. 3, p. 7.

Lillestøl, J. (1986). On the Problem of Optimal Timing of Slaughtering in Fish Farming. *Modeling, Identification and Control* 7: 199–207.

Lindbergh, J. (1987). The Economic Potential for the Commercial Production of Atlantic and Pacific Salmon in Chile. Paper presented at the Fundacion Chile conference on salmon aquaculture, Santiago, March 17–19, 1987.

Mendez, Z.R. (1987). Desarrollo y Estado de Situacion Actual de la Salmonicultura en Chile. Paper presented at the Fundacion Chile conference on salmon aquaculture, Santiago, March 17–19, 1987.

Mendez, Z.R. (1988). El Cultivo del Salmon. In *Chile Pesquero* No. 47, May-June 1988, pp. 27–30.

Noregs Offisielle Statistikk (1987). Fishing and Rearing of Salmon etc. Oslo: Central Bureau of Statistics of Norway.

Pierce, B. (1987). *Aquaculture in Alaska*. House Research Agency Report 87-B, Alaska State Legislature.

Reinert, A. (1987). Fish Farming Faroe Islands. Paper presented at F.E.S. Annual Assembly, Trondheim, May 8–11, 1987.

Ricker, W.E. (1975). *Computation and Interpretation of Biological Statistics in Fish Populations*. Ottawa: Environment Canada.

Ruckes, E. (1987). World Production and Salmon Markets: An Overview. Paper presented at the Fundacion Chile conference on salmon aquaculture (Santiago, March 17–19, 1987).

Salvanes, K. (1988). *Salmon Aquaculture in Norway: An empirical Analysis of cost and Production Properties*. Monograph No. 1, Institute of Fisheries Economics, Norwegian School of Economics and Business Administration. (In Norwegian.)

Shang, Y.C. (1981). *Aquaculture Economics: Basic Concepts and Methods of Analysis*. Boulder: Westview Press.

Shaw, S.A. (1989). *Markets for Farmed Salmon*. Report prepared for the FAO.

Shaw, S.A. and Muir, J.F. (1987). *Salmon: Economics and Marketing*. London: Croom Helm.

Shaw, S. and Rana, J. (1987). Market Outlet and Marketing Practices for Scottish Grown Salmon 1986. Market Report No. 3, Institute for Retail Studies, University of Stirling.

Singh, B. (1988). Market Structure and Price Formation: An Analysis of the United States Salmon Market. Working Paper No. 15/88, Centre for Applied Research, Norwegian School of Economics and Business Administration.

Todd (1987). Salmon Farming in New Zealand *Proceedings of the New Zealand Society of Animal Production* 47: 127–129.

Wurmann, C. (1985), Chilean Salmon: From Dreams to Reality. *Infofish Marketing Digest No. 4/85*.

Appendix 1
Optimal harvesting –
further analysis

In this appendix a number of extensions to the model for optimal harvesting are considered. In Chapter 3, the number of recruits was assumed to be given. Here release costs will be introduced into the model in order to analyse the optimal number of recruits to be released. Then insurance costs will be considered and the rotation problem analysed in greater detail. At the end, some mathematical derivations are shown.

For further readings, the interested reader is referred to Arnason (1987) and Bjørndal (1988).

Release costs

Release costs will now be introduced into the analysis. Let the cost per recruit be C_R and disregard other costs. The fish farmer's maximation problem then becomes:

$$\underset{\{0 \leqslant t \leqslant T,R\}}{\text{Max}} \quad \pi(R,t) = V(R,t)\,e^{-rt} - C_R R.$$

The biomass value is now a function of both the number of recruits and time. Thus the fish farmer has two control variables: number of recruits released (R) and harvesting time (t). The first order conditions for an optimum are:

(i)
$$\pi_r = \frac{\delta V}{\delta R}\,e^{-rt} - C_R = 0.$$

(ii)
$$\pi_t = \frac{\delta V}{\delta t}\,e^{-rt} - rV(R,t)\,e^{-rt} = 0.$$

Additionally the second order conditions must be satisfied. In this problem one will solve for the optimal number of recruits (R^*) and the optimal harvesting time (t^*) simultaneously, as biomass value is a function of both these variables. The first condition says that the discounted value of the marginal revenue with respect to recruits must equal the cost per recruit. Hence this condition decides how many recruits should be released, depending on the harvesting time. The second condition concerns determining the optimal harvesting time t^*, cf. Equation (6), which now also depends on the number of fish released (R).

Comparative statics can be used to analyse how changes in different parameter values affect harvesting time and the optimal number of recruits. First a change in the cost per recruit (C_R) will be considered. The analysis proceeds by totally differentiating the first order conditions:

$$\pi_{RR}\, dR + \pi_{Rt}\, dt = -\pi_{RC_R}\, dC_R$$
$$\pi_{tR}\, dR + \pi_{tt}\, dt = 0$$

By Cramer's rule the solutions can be found to be:

$$\frac{dR}{dC_R} = -\pi_{RC_R}\pi_{tt}/D,$$

$$\frac{dt}{dC_R} = -\pi_{RC_R}\pi_{tR}/D,$$

where $D = \pi_{RR}\pi_{tt} - \pi_{tR}2 > 0$. From the first order condition one obtains $\pi_{RC_R} = -1$, while π_{tt} is negative by the second order conditions. Accordingly, $\dfrac{dR}{dC_R}$ is negative. Furthermore,

$$\pi_{tR} = e^{-rt}\left[\frac{\delta V}{\delta t\, \delta R} - r\frac{\delta V}{\delta R}\right] = e^{-(r+M)t}\, p(w)\, w(t)\left[\frac{p'(w)}{p(w)}\, w'(t) + \frac{w'(t)}{w(t)} - M - r\right] = 0$$

Accordingly, $\dfrac{dt}{dC_R} = 0$.

The first result says that the optimal number of recruits is a decreasing function of the cost per recruit, which is intuitively obvious. The second result says that the optimal harvesting time is unaffected by changes in the cost per recruit.

To investigate the consequences of a change in the interest rate (r), one proceeds by differentiating the first order conditions:

$$\pi_{RR}\, dT + \pi_{Rt}\, dt = -\pi_{Rr}\, dr$$

$$\pi_{tR}\, dR + \pi_{tt}\, dt = -\pi_{tr}\, dr.$$

The following solutions are obtained:

$$\frac{dR}{dr} = -\pi_{Rt}\pi_{tt}/D$$

$$\frac{dt}{dr} = -\pi_{RR}\pi_{tr}/D$$

To evaluate the signs, note that

$$\pi_{Rt} = -r\frac{\delta V}{\delta R}e^{-rt} < 0 \text{ and } \pi_{tr} = -t\pi_t - V(R,t)e^{-rt} < 0,$$

so that both $\frac{dR}{dr}$ and $\frac{dt}{dr}$ are negative. These results say that an increase in the discount rate causes a decrease in both t^* and R^*.

Release costs introduced density-dependence in the value function. This could also have been done through the weight function:

$$w = w(N,t) = w(Re^{-Mt},t) = \tilde{w}(R,t).$$

Presumably growth will be decreasing (or more precisely, nonincreasing) in the number of fish released, while as before the fish grow over time. If this weight function is introduced in the value function, and the weight function is decreasing in the number of fish, one would purchase fewer recruits than otherwise.

Insurance

The cost of insuring the fish will be considered now. The other costs will be set equal to zero. The fish are insured according to their value at any time, i.e. $V(t)$. Let the insurance premium be k; e.g. $k = 0.04$ implies an insurance premium equal to four per cent of the value of the fish. As the fish are insured during their entire lifetime, total premium (P) constitutes

$$P = \int_0^t k\,V(u)\,du.$$

Insurance is a cost for the fish farmer. The maximation problem now is as follows:

$$\underset{0 \leqslant t \leqslant T}{\text{Max}}\ \pi(t) = V(t)e^{-rt} - \int_0^t k\,V(u)e^{-ru}\,du.$$

The first order condition for a maximum is given by:

$$\pi'(t) = V'(t) - rV(t) - kV(t) = 0.$$

The optimal harvesting time is thus given by

$$\frac{V'(t^*)}{V(t^*)} = r + k$$

which can be rewritten as:

$$\frac{p'(w)}{p(w)} w'(t^*) + \frac{w'(t^*)}{w(t^*)} = r + M + k.$$

This result shows that an insurance premium has the same effect on the optimal harvesting time as the discount rate and the natural mortality rate.

The rotation problem[1]

When rotation is considered (Section 3.4), the present value *(PV)* at time $t = 0$ of all future incomes is given by

$$NV = V(t_1)e^{-rt_1} + V(t_2)e^{-rt_2} + V(t_3)^{-rt_3} + \dots \qquad (A1)$$

The maximation of (A1) with respect to $t_1, t_2, t_3 \dots$ appears to be arduous. What simplifies the problem is that as all parameter values are constant over time, the rotation periods are of equal length, i.e.

$$t_k = k \cdot t, \qquad k = 1,2,3, \dots \qquad (A2)$$

where t is the rotation time. Inserting (A2) into (A1) gives the following maximation problem for the fish farmer (see summation formula below):

$$\text{Max } \pi(t) = V(t)e^{-rt} + V(t)e^{-2rt} + V(t)e^{-3rt} + V(t)e^{-4rt} + \dots$$

$$= \frac{V(t)}{e^{rt} - 1}$$

The first order condition for a maximum is given by

$$\pi'(t) = \frac{V'(t)[e^{rt} - 1] - re^{rt}V(t)}{[e^{rt} - 1]^2} = 0.$$

This can be simplified to

$$\frac{V'(t^*)}{V(t^*)} = \frac{r}{1 - e^{-rt^*}} \qquad (A3)$$

t^* is now to be understood as the *optimal rotation time*. Equation (A3) is known as the Faustmann rule after the German forester Martin Faustmann, who was the first to solve the optimal rotation problem in forestry. By comparing this expression with the formula for optimal harvesting time for a single yearclass (6'), it is seen that the left hand side is the same, while there is a new component on the right hand side. As $(1-e^{-rt}) < 1$, rotation implies a reduction in the optimal harvesting time compared to no rotation.

Equation (A3) can be rewritten as (see below):

$$V'(t^*) = rV(t^*) + r\,\frac{V(t^*)}{e^{rt^*} - 1} \qquad\qquad (A3')$$

As before $V'(t)$ is the change in biomass value over time, whereas $rV(t)$ is the opportunity cost of the biomass value. The expression

$$\frac{V(t)}{e^{rt}-1}$$

is the present value of future cash flows and is the component which takes the rotation aspect into account. It gives the value of the fish farm which, multiplied by the interest rate, determines the *opportunity cost of the fish farm*. This value is a result of the fact that farm size is limited.

A special case occurs when the interest rate is zero. By use of l'Hôpital's rule (see below), the right hand side of Equation (A3) becomes:

$$\lim_{r \Rightarrow 0}\frac{r}{1-e^{-rt}} = \frac{1}{t}$$

so that Equation (A3) becomes

$$\frac{V'(t^*)}{V(t^*)} = \frac{1}{t^*}, \qquad\qquad r = 0 \qquad\qquad (A4)$$

or

$$\frac{V(t^*)}{t^*} = V'(t^*), \qquad\qquad r = 0. \qquad\qquad (A4')$$

In this case fish should be harvested when the time change in the biomass value $(V'(t))$ equals the average biomass value $(V(t)/t)$. This means that one maximizes annual biomass value $V(t)/t$.

Mathematical derivations

The relationship between continuous time and discrete time interest rates

Continuous time interest rate $= r$
Discrete time interest rate $= i$

There is the following relationship between the two interest rates:

$$(1 + i)^t = e^{rt}$$

Series summation

Let (1) $S = e^{-rt} + e^{-2rt} + e^{-3rt} + \dots$

Multiply both sides with e^{rt}:

(2) $e^{rt}S = 1 + e^{-rt} + e^{-2rt} + e^{-3rt} + \dots$

Subtract (1) from (2):

$e^{rt}S - S = 1$

$\Rightarrow \quad S = \dfrac{1}{e^{rt} - 1}$

Rewriting of equation (A3)

$$\frac{V'(t^*)}{V(t^*)} = \frac{r}{1 - e^{-rt^*}}$$

Multiply the numerator and the denominator on the right side with e^{rt^*}:

$$V'(t^*) = \frac{e^{rt^*} rV(t^*)}{e^{rt^*} - 1}$$

Add and subtract $rV(t^*)$ in the denominator on the right hand side:

$$V'(t^*) = \frac{rV(t^*)(e^{rt^*} - 1) + rV(t^*)}{e^{rt^*} - 1}$$

$$V'(t) = rV(t^*) + r\frac{V(t^*)}{e^{rt^*} - 1}$$

Use of l'Hôpital's rule

$$\lim_{r \to 0} \frac{r}{1 - e^{rt}} = \lim_{r \to 0} \frac{\frac{d}{dr}(r)}{\frac{d}{dr}[1 - e^{-rt}]} = \lim_{r \to 0} \frac{1}{te^{-rt}} = \frac{1}{te^{-0}} = \frac{1}{t}$$

l'Hôpital's rule is used to evaluate the value of an expression when both the numerator and the denominator approach zero as a result of a variable approaching a certain value; in this case, r approaching zero.

NOTE

1 This section is based in part on Clark (1976), chapter 8.

Appendix 2
Statistical tables

Table A1   Wild Salmon Landings by Species 1980–1985 ('000 tonnes)

Year	Atlantic	Chinook	Coho	Sockeye	Pink	Chum	Cherry	Total
1980	12.0	22.0	31.9	111.8	226.1	166.8	2.8	573.4
1981	12.0	22.8	28.5	132.7	264.9	184.8	3.3	649.0
1982	9.3	25.1	41.6	128.2	170.4	178.9	3.7	557.2
1983	11.3	18.9	29.6	163.8	255.1	196.0	4.0	678.7
1984	12.4	18.4	41.0	126.8	210.5	211.0	4.0	624.1
1985	10.1	20.1	39.4	150.8	301.1	268.0	4.0	793.5

Source: Food and Agriculture Organisation of the United Nations, Rome, Italy. *Yearbook of Fishery Statistics*

Table A2   Production and Landings of Atlantic, Chinook and Coho Salmon ('000 tonnes)

	Landings of wild Atlantic, coho and chinook	Production of farmed Atlantic, coho and chinook	Farmed as a per cent of total	Total
1980	65.9	4.8	7	70.7
1981	63.3	11.6	15	74.9
1982	76.0	16.5	18	92.5
1983	59.8	24.6	29	84.4
1984	71.8	32.6	31	104.4
1985	69.6	47.1	40	116.7

Table A3   Production and Landings of Atlantic, Chinook, Coho and Sockeye Salmon, 1980–1985 ('000 tonnes)

	Landings of wild Atlantic, coho and chinook and sockeye	Production of farmed Atlantic, coho and chinook	Farmed as a per cent of total	Total
1980	177.7	4.8	3	182.5
1981	196.0	11.6	6	207.6
1982	204.2	16.5	7	220.7
1983	223.6	24.6	10	248.2
1984	198.6	32.6	14	231.2
1985	220.4	47.1	18	267.5

Table A4   Wild Salmon Landings 1981–1988 by Country ('000 tonnes)

Year	USA	Japan	USSR	Canada	Others	Total
1981	293.6	158.3	108.8	80.7	7.6	649.0
1982	274.8	144.4	67.5	66.6	3.9	557.2
1983	289.1	170.9	134.1	75.3	9.3	678.7
1984	312.4	167.8	81.8	50.8	11.3	624.1
1985	328.4	215.9	132.4	108.0	8.8	793.5
1986[a]	299.0	186.0	80.0	101.0	9.0	675.0
1987[a]	255.2	180.0	139.0	66.0	10.0	650.2
1988[a]	–	–	–	80.0	–	–
Per cent of catch						
1981–1987	44	27	16	12	1	100

[a]   Preliminary figure.

Sources: Food and Agriculture Organisation of the United Nations, Rome, Italy. *Yearbook of Fishery Statistics* for 1981–1985

1986–1987: Fisheries of the United States 1986–87, U.S. Department of Commerce for United States. FAO/GLOBEFISH Highlights (4/88) for other countries.

1988: Leitz (1989) for Canada.

Table A5   Weight Observations for Salmon

Time[a] (year)	Weight (kg)
0.1	0.15
0.56	0.61
1.292	2.2
1.458	3.6
1.708	4.7
2.042	4.95
2.208	4.9

[a] Time measured in years after release.

Source: Fishery biologist Thore Thomassen.

Table A6   Weight Observations for Turbot

Time (year)	Weight (kg)
0.125	0.05
0.5	0.175
1.0	0.5
1.5	1.0375
2.0	1.5875
2.5	2.1
2.875	2.2

The data are from Golden Sea Produce's land-based turbot farm in Scotland.

Index

Atlantic Canada *see* Eastern Canada
Atlantic salmon 1, 3, 4, 7, 9, 18, 19, 20, 21, 22, 25, 29, 30, 33, 34, 94, 95, 97, 98

biology 1, 2
biomass 41, 42, 43, 44, 45, 46, 59, 61, 68, 69, 70, 73, 75, 76, 77, 78
British Columbia 8, 17, 18, 19, 20, 37, 80, 94, 95, 96, 97, 98, 99

Canada 7, 8, 9, 18, 21, 67
    see also British Columbia and Eastern Canada
capture fisheries 1, 4, 5, 6
cash flow 42, 64, 65, 66, 67, 74, 75, 76, 100, 102, 104, 105
cherry salmon 6
Chile 7, 8, 9, 21, 22, 26, 35, 37
chinook salmon 3, 4, 6, 7, 8, 9, 17, 18, 19, 21, 26, 30, 67, 94, 95, 96, 98
chum salmon 6, 30
coho salmon 4, 6, 7, 8, 9, 16, 17, 18, 19, 21, 25, 26, 30, 94, 95
conversion ratio 51, 59, 61, 71, 73, 82, 88, 90, 91, 92, 96, 98, 105

Denmark 30, 31, 32

Eastern Canada 8, 20
environment 1, 2, 5, 19, 36
environmental 1, 3, 9, 12, 25
European Economic Community 15, 30, 32, 36

facility investments 81, 87, 95, 100, 103
Faroe Islands 8, 23
France 23, 28, 29, 30, 31, 32, 33
fry 3, 4, 53, 57

hatchery 3, 16, 18, 20, 23, 26, 27, 80, 81, 82

Iceland 8, 24
investment costs 5, 81, 86, 87
Ireland 7, 8, 22, 23, 36, 37

Japan 7, 8, 9, 16, 17, 26, 28, 29, 30, 31, 32, 35

land-based farm 2, 57

New Zealand 8, 26
Norway 5, 7, 8, 9, 10, 11, 12, 13, 14, 15, 17, 18, 19, 23, 29, 30, 31, 37, 80, 81, 87, 95, 98

ocean ranching 2, 21, 24, 48, 58

Pacific salmon 1, 3, 4, 9, 16, 17, 18, 20, 25, 26, 29, 30, 37, 95, 98
pens *see* sea-pens
pink salmon 6
platform installation 4
production planning 64, 77

rotation 55, 56, 57, 61, 65

Scotland 7, 8, 14, 15, 16, 22, 26, 29, 35, 36, 37
sea-pens 2, 18
sexual maturation 7, 76
sockeye salmon 6, 7, 8, 9, 30
Spain 31
spawning 3, 4, 6, 57

technology 1, 2, 4, 9, 15, 35, 36, 100

United Kingdom 7, 15, 23, 29, 31
USA 7, 8, 15, 24, 25, 28, 29, 30, 31, 32, 33, 34, 36, 37

West Germany 30, 31, 32
working capital 81, 88, 96, 101

Books published by
Fishing News Books

Free catalogue available on request

Advances in fish science and technology
Aquaculture in Taiwan
Aquaculture: principles and practice
Aquaculture training manual
Aquatic weed control
Atlantic salmon: its future
Better angling with simple science
British freshwater fishes
Business management in fisheries and
  aquaculture
Cage aquaculture
Calculations for fishing gear designs
Carp farming
Commercial fishing methods
Control of fish quality
Crab and lobster fishing
The crayfish
Culture of bivalve molluscs
Design of small fishing vessels
Developments in electric fishing
Developments in fisheries research in
  Scotland
Echo sounding and sonar for fishing
The economics of salmon aquaculture
The edible crab and its fishery in British
  waters
Eel culture
Engineering, economics and fisheries
  management
European inland water fish: a multilingual
  catalogue
FAO catalogue of fishing gear designs
FAO catalogue of small scale fishing gear
Fibre ropes for fishing gear
Fish and shellfish farming in coastal waters
Fish catching methods of the world
Fisheries oceanography and ecology
Fisheries of Australia
Fisheries sonar
Fisherman's workbook
Fishermen's handbook
Fishery development experiences
Fishing and stock fluctuations
Fishing boats and their equipment
Fishing boats of the world 1
Fishing boats of the world 2
Fishing boats of the world 3
The fishing cadet's handbook
Fishing ports and markets
Fishing with electricity

Fishing with light
Freezing and irradiation of fish
Freshwater fisheries management
Glossary of UK fishing gear terms
Handbook of trout and salmon diseases
A history of marine fish culture in Europe and
  North America
How to make and set nets
Inland aquaculture development handbook
Intensive fish farming
Introduction to fishery by-products
The law of aquaculture: the law relating to the
  farming of fish and shellfish in Great Britain
The lemon sole
A living from lobsters
The mackerel
Making and managing a trout lake
Managerial effectiveness in fisheries and
  aquaculture
Marine fisheries ecosystem
Marine pollution and sea life
Marketing in fisheries and aquaculture
Mending of fishing nets
Modern deep sea trawling gear
More Scottish fishing craft
Multilingual dictionary of fish and fish products
Navigation primer for fishermen
Net work exercises
Netting materials for fishing gear
Ocean forum
Pair trawling and pair seining
Pelagic and semi-pelagic trawling gear
Penaeid shrimps — their biology and
  management
Planning of aquaculture development
Refrigeration of fishing vessels
Salmon and trout farming in Norway
Salmon farming handbook
Scallop and queen fisheries in the British Isles
Seine fishing
Squid jigging from small boats
Stability and trim of fishing vessels and other
  small ships
Study of the sea
Textbook of fish culture
Training fishermen at sea
Trends in fish utilization
Trout farming handbook
Trout farming manual
Tuna fishing with pole and line